Numerical Analysis or Numerical Method in Symmetry

Numerical Analysis or Numerical Method in Symmetry

Special Issue Editor

Clemente Cesarano

MDPI • Basel • Beijing • Wuhan • Barcelona • Belgrade • Manchester • Tokyo • Cluj • Tianjin

Special Issue Editor
Clemente Cesarano
Uninettuno University
Italy

Editorial Office
MDPI
St. Alban-Anlage 66
4052 Basel, Switzerland

This is a reprint of articles from the Special Issue published online in the open access journal *Symmetry* (ISSN 2073-8994) (available at: https://www.mdpi.com/journal/symmetry/special_issues/Numerical_Analysis).

For citation purposes, cite each article independently as indicated on the article page online and as indicated below:

LastName, A.A.; LastName, B.B.; LastName, C.C. Article Title. *Journal Name* **Year**, *Article Number*, Page Range.

ISBN 978-3-03928-372-9 (Pbk)
ISBN 978-3-03928-373-6 (PDF)

© 2020 by the authors. Articles in this book are Open Access and distributed under the Creative Commons Attribution (CC BY) license, which allows users to download, copy and build upon published articles, as long as the author and publisher are properly credited, which ensures maximum dissemination and a wider impact of our publications.

The book as a whole is distributed by MDPI under the terms and conditions of the Creative Commons license CC BY-NC-ND.

Contents

About the Special Issue Editor vii

Preface to "Numerical Analysis or Numerical Method in Symmetry" ix

Hyun Geun Lee
Numerical Simulation of Pattern Formation on Surfaces Using an Efficient Linear
Second-Order Method
Reprinted from: *Symmetry* **2019**, *11*, 1010, doi:10.3390/sym11081010 1

Mutaz Mohammad
A Numerical Solution of Fredholm Integral Equations of the Second Kind Based on Tight
Framelets Generated by the Oblique Extension Principle
Reprinted from: *Symmetry* **2019**, *11*, 854, doi:10.3390/sym11070854 9

Clemente Cesarano and Omar Bazighifan
Qualitative Behavior of Solutions of Second Order Differential Equations
Reprinted from: *Symmetry* **2019**, *11*, 777, doi:10.3390/sym11060777 25

SAIRA and Shuhuang Xiang
Approximation to Logarithmic-Cauchy Type Singular Integrals with Highly Oscillatory Kernels
Reprinted from: *Symmetry* **2019**, *11*, 728, doi:10.3390/sym11060728 33

Dario Assante and Luigi Verolino
Model for the Evaluation of an Angular Slot's Coupling Impedance
Reprinted from: *Symmetry* **2019**, *11*, 700, doi:10.3390/sym11050700 47

Clemente Cesarano and Omar Bazighifan
Asymptotic Properties of Solutions of Fourth-Order Delay Differential Equations
Reprinted from: *Symmetry* **2019**, *11*, 628, doi:10.3390/sym11050628 59

Clemente Cesarano, Sandra Pinelas and Paolo Emilio Ricci
The Third and Fourth Kind Pseudo-Chebyshev Polynomials of Half-Integer Degree
Reprinted from: *Symmetry* **2019**, *11*, 274, doi:10.3390/sym11020274 69

Siti Nur Alwani Salleh, Norfifah Bachok, Norihan Md Arifin and Fadzilah Md Ali
Numerical Analysis of Boundary Layer Flow Adjacent to a Thin Needle in Nanofluid with the
Presence of Heat Source and Chemical Reaction
Reprinted from: *Symmetry* **2019**, *11*, 543, doi:10.3390/sym11040543 81

Shengfeng Li and Yi Dong
k-Hypergeometric Series Solutions to One Type of Non-Homogeneous
k-Hypergeometric Equations
Reprinted from: *Symmetry* **2019**, *11*, 262, doi:10.3390/sym11020262 97

**Junaid Ahmad, Yousaf Habib, Azqa Ashraf, Saba Shafiq, Muhammad Younas and
Shafiq ur Rehman**
Symplectic Effective Order Numerical Methods for Separable Hamiltonian Systems
Reprinted from: *Symmetry* **2019**, *11*, 142, doi:10.3390/sym11020142 109

Nizam Ghawadri, Norazak Senu, Firas Adel Fawzi, Fudziah Ismail and Zarina Bibi Ibrahim
Explicit Integrator of Runge-Kutta Type for Direct Solution of $u^{(4)} = f(x, u, u', u'')$
Reprinted from: *Symmetry* **2019**, *11*, 246, doi:10.3390/sym11020246 123

Yun-Jeong Cho, Kichang Im, Dongkoo Shon, Daehoon Park and Jong-Myon Kim
Improvement of Risk Assessment Using Numerical Analysis for an Offshore Plant Dipole Antenna
Reprinted from: *Symmetry* **2018**, *10*, 681, doi:10.3390/sym10120681 . . . . **153**

Wajeeha Irshad, Yousaf Habib and Muhammad Umar Farooq
A Complex Lie-Symmetry Approach to Calculate First Integrals and Their Numerical Preservation
Reprinted from: *Symmetry* **2019**, *11*, 11, doi:10.3390/sym11010011 . . . . **167**

About the Special Issue Editor

Clemente Cesarano is associate professor of Numerical Analysis at the Section of Mathematics -Uninettuno University, Rome Italy; he is the coordinator of the doctoral college in Technological Innovation Engineering, coordinator of the Section of Mathematics, vice-dean of the Faculty of Engineering, president of the Degree Course in Management Engineering, director of the Master in Project Management Techniques, and coordinator of the Master in Applied and Industrial Mathematics. He is also a member of the Research Project "Modeling and Simulation of the Fractionary and Medical Center", Complutense University of Madrid (Spain) and head of the national group from 2015, member of the Research Project (Serbian Ministry of Education and Science) "Approximation of Integral and Differential Operators and Applications", University of Belgrade (Serbia) and coordinator of the national group from 2011-), a member of the Doctoral College in Mathematics at the Department of Mathematics of the University of Mazandaran (Iran), expert (Reprise) at the Ministry of Education, University and Research, for the ERC sectors: Analysis, Operator algebras and functional analysis, Numerical analysis. Clemente Cesarano is Honorary Fellows of the Australian Institute of High Energetic Materials, affiliated with the National Institute of High Mathematics (INdAM), is affiliated with the International Research Center for the "Mathematics & Mechanics of Complex Systems" (MEMOCS) - University of L'Aquila, associate of the CNR at the Institute of Complex Systems (ISC), affiliated with the "Research ITalian network on Approximation (RITA)" network as the head of the Uninettuno office, UMI member, SIMAI member.

Preface to "Numerical Analysis or Numerical Method in Symmetry"

Numerical methods and, in particular, numerical analysis represent important fields of investigation in modern mathematical research. In recent years, numerical analysis has undertaken various lines of application in different areas of applied mathematics and, moreover, in applied sciences, such as biology, physics, engineering, and so on. However, part of the research on the topic of numerical analysis cannot exclude the fundamental role played by approximation theory and some of the tools used to develop this research. In this Special Issue, we want to draw attention to mathematical methods used in numerical analysis, such as special functions, orthogonal polynomials and their theoretical instruments, such as Lie algebra, to investigate the concepts and properties of some special and advanced methods that are useful in the description of solutions of linear and non-linear differential equations. A further field of investigation is devoted to the theory and related properties of fractional calculus with its suitable application to numerical methods.

Clemente Cesarano
Special Issue Editor

Article

Numerical Simulation of Pattern Formation on Surfaces Using an Efficient Linear Second-Order Method

Hyun Geun Lee

Department of Mathematics, Kwangwoon University, Seoul 01897, Korea; leeh1@kw.ac.kr

Received: 4 June 2019; Accepted: 25 July 2019; Published: 5 August 2019

Abstract: We present an efficient linear second-order method for a Swift–Hohenberg (SH) type of a partial differential equation having quadratic-cubic nonlinearity on surfaces to simulate pattern formation on surfaces numerically. The equation is symmetric under a change of sign of the density field if there is no quadratic nonlinearity. We introduce a narrow band neighborhood of a surface and extend the equation on the surface to the narrow band domain. By applying a pseudo-Neumann boundary condition through the closest point, the Laplace–Beltrami operator can be replaced by the standard Laplacian operator. The equation on the narrow band domain is split into one linear and two nonlinear subequations, where the nonlinear subequations are independent of spatial derivatives and thus are ordinary differential equations and have closed-form solutions. Therefore, we only solve the linear subequation on the narrow band domain using the Crank–Nicolson method. Numerical experiments on various surfaces are given verifying the accuracy and efficiency of the proposed method.

Keywords: Swift–Hohenberg type of equation; surfaces; narrow band domain; closest point method; operator splitting method

1. Introduction

A Swift–Hohenberg (SH) type of partial differential equation [1] has been used to study pattern formation [2–5]:

$$\frac{\partial \phi}{\partial t} = -\left(\phi^3 - g\phi^2 + \left(-\epsilon + (1+\Delta)^2\right)\phi\right),$$

where ϕ is the density field and $g \geq 0$ and $\epsilon > 0$ are constants. In general, the equation does not have an analytical solution, thus various computational algorithms [6–13] have been proposed to obtain a numerical solution. However, most of them were solved on flat surfaces except [12,13].

In this paper, we present an efficient linear second-order method for the SH type of equation on surfaces, which is based on the closest point method [14,15]. We introduce a narrow band domain of a surface and apply a pseudo-Neumann boundary condition on the boundary of the narrow band domain through the closest point [16]. This results in a constant value of ϕ in the direction normal to the surface, thus the Laplace–Beltrami operator can be replaced by the standard Laplacian operator. In addition, we split the equation into one linear and two nonlinear subequations [17,18], where the nonlinear subequations are independent of spatial derivatives and thus are ordinary differential equations and have closed-form solutions. Therefore, we only solve the linear subequation on the narrow band domain using the Crank–Nicolson method. As a result, our method is easy to implement and linear.

This paper is organized as follows. In Section 2, we describe the SH type of equation on a narrow band domain. In Section 3, we propose an efficient linear second-order method for the equation on the narrow band domain. Numerical examples on various surfaces are given in Section 4. Finally, we conclude in Section 5.

2. Swift–Hohenberg Type of Equation on a Narrow Band Domain

The SH type of equation on a surface $\mathcal{S}$ is given by

$$\frac{\partial \phi(\mathbf{x},t)}{\partial t} = -\left(\phi^3(\mathbf{x},t) - g\phi^2(\mathbf{x},t) + \left(-\epsilon + (1+\Delta_{\mathcal{S}})^2\right)\phi(\mathbf{x},t)\right), \quad \mathbf{x} \in \mathcal{S}, \ 0 < t \leq T, \tag{1}$$

where $\Delta_{\mathcal{S}}$ is the Laplace–Beltrami operator [19,20]. Next, let $\Omega_\delta = \{\mathbf{y} \mid \mathbf{x} \in \mathcal{S}, \mathbf{y} = \mathbf{x} + \eta \mathbf{n}(\mathbf{x})$ for $|\eta| < \delta\}$ be a δ-neighborhood of $\mathcal{S}$, where $\mathbf{n}(\mathbf{x})$ is a unit normal vector at $\mathbf{x}$. Then, we extend the Equation (1) to the narrow band domain Ω_δ:

$$\frac{\partial \phi(\mathbf{x},t)}{\partial t} = -\left(\phi^3(\mathbf{x},t) - g\phi^2(\mathbf{x},t) + \left(-\epsilon + (1+\Delta_{\mathcal{S}})^2\right)\phi(\mathbf{x},t)\right), \quad \mathbf{x} \in \Omega_\delta, \ 0 < t \leq T \tag{2}$$

with the pseudo-Neumann boundary condition on $\partial\Omega_\delta$:

$$\phi(\mathbf{x},t) = \phi(\mathrm{cp}(\mathbf{x}),t), \tag{3}$$

where $\mathrm{cp}(\mathbf{x})$ is a point on $\mathcal{S}$, which is closest to $\mathbf{x} \in \partial\Omega_\delta$ [14]. For a sufficiently small δ, ϕ is constant in the direction normal to the surface. Thus, the Laplace–Beltrami operator in Ω_δ can be replaced by the standard Laplacian operator [14], i.e.,

$$\frac{\partial \phi(\mathbf{x},t)}{\partial t} = -\left(\phi^3(\mathbf{x},t) - g\phi^2(\mathbf{x},t) + \left(-\epsilon + (1+\Delta)^2\right)\phi(\mathbf{x},t)\right), \quad \mathbf{x} \in \Omega_\delta, \ 0 < t \leq T. \tag{4}$$

3. Numerical Method

In this section, we propose an efficient linear second-order method for solving Equation (4) with the boundary condition (3). We discretize Equation (4) in $\Omega = [-L_x/2, L_x/2] \times [-L_y/2, L_y/2] \times [-L_z/2, L_z/2]$ that includes Ω_δ. Let $h = L_x/N_x = L_y/N_y = L_z/N_z$ be the uniform grid size, where N_x, N_y, and N_z are positive integers. Let $\Omega^h = \{\mathbf{x}_{ijk} = (x_i, y_j, z_k) \mid x_i = -L_x/2 + ih, y_j = -L_y/2 + jh, z_k = -L_z/2 + kh$ for $0 \leq i \leq N_x, 0 \leq j \leq N_y, 0 \leq k \leq N_z\}$ be a discrete domain. Let ϕ_{ijk}^n be an approximation of $\phi(\mathbf{x}_{ijk}, n\Delta t)$, where Δt is the time step. Let $\Omega_\delta^h = \{\mathbf{x}_{ijk} \mid |\psi_{ijk}| < \delta\}$ be a discrete narrow band domain, where ψ is a signed distance function for the surface $\mathcal{S}$, and $\partial\Omega_\delta^h = \{\mathbf{x}_{ijk} \mid I_{ijk} |\nabla_h I_{ijk}| \neq 0\}$ are discrete domain boundary points, where $\nabla_h I_{ijk} = (I_{i+1,j,k} - I_{i-1,j,k}, I_{i,j+1,k} - I_{i,j-1,k}, I_{i,j,k+1} - I_{i,j,k-1})/(2h)$. Here, $I_{ijk} = 0$ if $\mathbf{x}_{ijk} \in \Omega_\delta^h$, and $I_{ijk} = 1$, otherwise.

We here split Equation (4) into the following subequations:

$$\frac{\partial \phi}{\partial t} = -(\phi^3 - \epsilon\phi), \tag{5}$$

$$\frac{\partial \phi}{\partial t} = g\phi^2, \tag{6}$$

$$\frac{\partial \phi}{\partial t} = -(1+\Delta)^2\phi. \tag{7}$$

Equations (5) and (6) are solved analytically and the solutions ϕ_{ijk}^{n+1} are given as follows:

$$\phi_{ijk}^{n+1} = \frac{\phi_{ijk}^n}{\sqrt{(\phi_{ijk}^n)^2/\epsilon + (1 - (\phi_{ijk}^n)^2/\epsilon)e^{-2\epsilon\Delta t}}} \quad \text{and} \quad \phi_{ijk}^{n+1} = \frac{\phi_{ijk}^n}{1 - g\Delta t \phi_{ijk}^n},$$

respectively. In addition, Equation (7) is solved using the Crank–Nicolson method:

$$\frac{\phi_{ijk}^{n+1} - \phi_{ijk}^n}{\Delta t} = -\frac{(1+\Delta_h)^2}{2}(\phi_{ijk}^{n+1} + \phi_{ijk}^n) \qquad (8)$$

with the boundary condition on $\partial \Omega_\delta^h$:

$$\phi_{ijk}^n = \phi^n(\text{cp}(\mathbf{x}_{ijk})).$$

Here, $\Delta_h \phi_{ijk} = (\phi_{i+1,j,k} + \phi_{i-1,j,k} + \phi_{i,j+1,k} + \phi_{i,j-1,k} + \phi_{i,j,k+1} + \phi_{i,j,k-1} - 6\phi_{ijk})/h^2$. The numerical closest point $\text{cp}(\mathbf{x}_{ijk})$ for a point $\mathbf{x}_{ijk} \in \partial \Omega_\delta^h$ is defined as

$$\text{cp}(\mathbf{x}_{ijk}) = \mathbf{x}_{ijk} - |\psi_{ijk}| \frac{\nabla_h |\psi_{ijk}|}{|\nabla_h |\psi_{ijk}||}.$$

In general, $\text{cp}(\mathbf{x}_{ijk})$ is not a grid point in Ω_δ^h, i.e., $\text{cp}(\mathbf{x}_{ijk}) \notin \Omega_\delta^h$, and thus we use trilinear interpolation and take $\delta > \sqrt{3}h$ to obtain $\phi(\text{cp}(\mathbf{x}_{ijk}))$. The resulting implicit linear discrete system of Equation (8) is solved efficiently using the Jacobi iterative method. We iterate the Jacobi iteration until a discrete L^2-norm of the consecutive error on Ω_δ^h is less than a tolerance *tol*. Here, the discrete L^2-norm on Ω_δ^h is defined as $\|\phi\|_{L^2(\Omega_\delta^h)} = \sqrt{\sum_{\mathbf{x}_{ijk} \in \Omega_\delta^h} \phi_{ijk}^2 / \#\Omega_\delta^h}$, where $\#\Omega_\delta^h$ is the cardinality of Ω_δ^h. Then, the second-order solution of Equation (4) is evolved by five stages [21]

$$\phi_{ijk}^{(1)} = \frac{\phi_{ijk}^n}{\sqrt{(\phi_{ijk}^n)^2/\epsilon + (1 - (\phi_{ijk}^n)^2/\epsilon)e^{-\epsilon \Delta t}}},$$

$$\phi_{ijk}^{(2)} = \frac{\phi_{ijk}^{(1)}}{1 - (g\Delta t/2)\phi_{ijk}^{(1)}},$$

$$\frac{\phi_{ijk}^{(3)} - \phi_{ijk}^{(2)}}{\Delta t} = -\frac{(1+\Delta_h)^2}{2}(\phi_{ijk}^{(3)} + \phi_{ijk}^{(2)}),$$

$$\phi_{ijk}^{(4)} = \frac{\phi_{ijk}^{(3)}}{1 - (g\Delta t/2)\phi_{ijk}^{(3)}},$$

$$\phi_{ijk}^{n+1} = \frac{\phi_{ijk}^{(4)}}{\sqrt{(\phi_{ijk}^{(4)})^2/\epsilon + (1 - (\phi_{ijk}^{(4)})^2/\epsilon)e^{-\epsilon \Delta t}}}.$$

4. Numerical Experiments

4.1. Convergence Test

In order to verify the rate of convergence of the proposed method, we consider the evolution of ϕ on a unit sphere. An initial piece of data is

$$\phi(x,y,z,0) = 0.15 + 0.1\cos(2\pi x)\cos(2\pi y)\cos(2\pi z)$$

and a signed distance function for the unit sphere is

$$\psi(x,y,z) = \sqrt{x^2 + y^2 + z^2} - 1.$$

on $\Omega = [-1.5, 1.5]^3$. We fix the grid size to $h = 0.125$ and vary $\Delta t = T/2, T/2^2, T/2^3, T/2^4$ for $T = 0.00025$ with $\epsilon = 0.25$, $\delta = 2.2\sqrt{3}h$, and $tol = \Delta t$. Table 1 shows the L^2-errors of $\phi(x, y, z, T)$ and convergence rates with $g = 0$. Here, the errors are computed by comparison with a reference numerical solution using $\Delta t = T/2^6$. It is observed that the method is second-order accurate in time. Note that we obtain the same result for $g = 1$.

Table 1. L^2-errors and convergence rates for $g = 0$.

Δt	$T/2$		$T/2^2$		$T/2^3$		$T/2^4$
L^2-error	5.445×10^{-3}		1.341×10^{-3}		2.862×10^{-4}		5.519×10^{-5}
Rate		2.02		2.22		2.37	

4.2. Pattern Formation on a Sphere

Unless otherwise stated, we take an initial piece of data as

$$\phi(x, y, z, 0) = 0.15 + \text{rand}(x, y, z),$$

where $\text{rand}(x, y, z)$ is a uniformly distributed random number between -0.1 and 0.1 at the grid points, and use $\epsilon = 0.25$, $h = 1$, $\Delta t = 0.1$, $\delta = 1.1\sqrt{3}h$, and $tol = 10^{-4}$.

For $g = 0$ and 1, Figures 1 and 2 show the evolution of $\phi(x, y, z, t)$ on a sphere with $\psi(x, y, z) = \sqrt{x^2 + y^2 + z^2} - 32$ on $\Omega = [-36, 36]^3$, respectively. Depending on the value of g, we have different patterns, such as striped (Figure 1) and hexagonal (Figure 2) [11]. Figure 3 shows the energy decay with $g = 0$ and 1, where the energy $\mathcal{E}(\phi)$ is defined by

$$\mathcal{E}(\phi) = \int_{\Omega_\delta} \left(\frac{1}{4}\phi^4 - \frac{g}{3}\phi^3 + \frac{1}{2}\phi \left(-\epsilon + (1 + \Delta)^2 \right) \phi \right) dx.$$

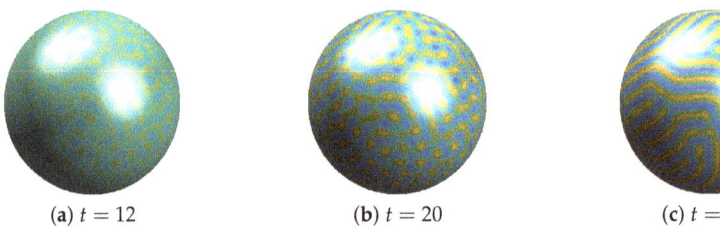

(a) $t = 12$ (b) $t = 20$ (c) $t = 100$

Figure 1. Evolution of $\phi(x, y, z, t)$ with $g = 0$. The yellow and blue regions indicate $\phi = 0.7540$ and -0.7783, respectively.

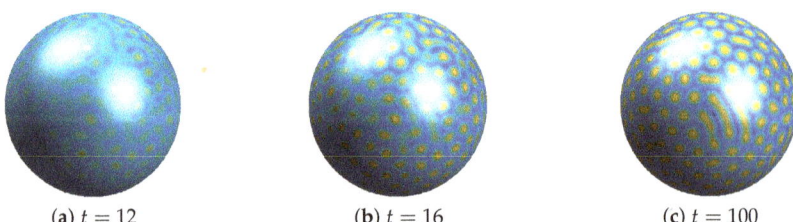

(a) $t = 12$ (b) $t = 16$ (c) $t = 100$

Figure 2. Evolution of $\phi(x, y, z, t)$ with $g = 1$. The yellow and blue regions indicate $\phi = 1.4320$ and -0.7152, respectively.

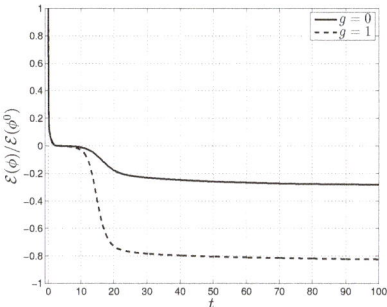

Figure 3. Evolution of $\mathcal{E}(\phi)/\mathcal{E}(\phi^0)$ on the sphere with $g = 0$ and 1.

4.3. Pattern Formation on a Sphere Perturbed by a Spherical Harmonic

In this section, we perform the evolution of ϕ on a sphere of center $(0,0,0)$ and radius 32 perturbed by a spherical harmonic $10\, Y_{10}^7(\theta, \varphi)$. Here, θ and φ are the polar and azimuthal angles, respectively, and the computational domain is $\Omega = [-40, 40]^3$. Figures 4 and 5 show the evolution of $\phi(x, y, z, t)$ with $g = 0$ and 1, respectively. From the results in Figures 4 and 5, we can see that our method can solve the SH type of equation on not only simple but also complex surfaces. Figure 6 shows the energy decay with $g = 0$ and 1.

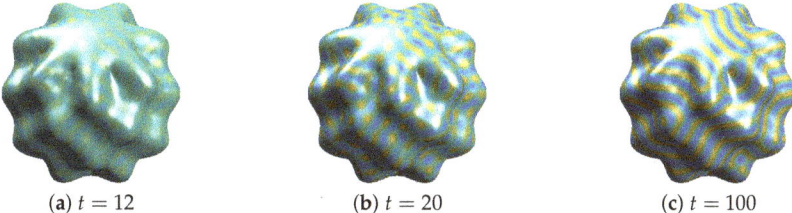

(a) $t = 12$ (b) $t = 20$ (c) $t = 100$

Figure 4. Evolution of $\phi(x, y, z, t)$ with $g = 0$. The yellow and blue regions indicate $\phi = 0.8717$ and -0.8372, respectively.

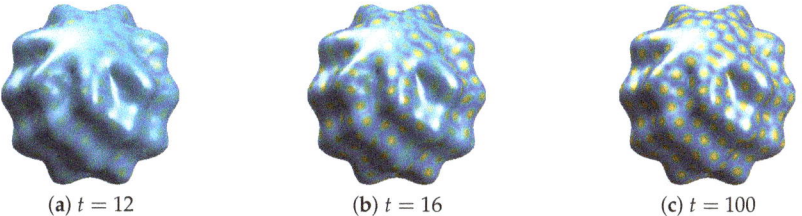

(a) $t = 12$ (b) $t = 16$ (c) $t = 100$

Figure 5. Evolution of $\phi(x, y, z, t)$ with $g = 1$. The yellow and blue regions indicate $\phi = 1.4833$ and -0.7135, respectively.

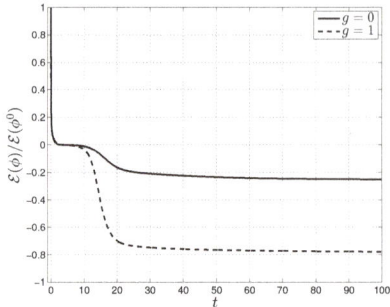

Figure 6. Evolution of $\mathcal{E}(\phi)/\mathcal{E}(\phi^0)$ on the perturbed sphere with $g = 0$ and 1.

4.4. Pattern Formation on a Spindle

Finally, we simulate the evolution of ϕ on a spindle that has narrow and sharp tips. The spindle is defined parametrically as

$$x = 16\cos\theta \sin\varphi, \quad y = 16\sin\theta \sin\varphi, \quad z = 32\left(\frac{2\varphi}{\pi} - 1\right),$$

where $\theta \in [0, 2\pi)$ and $\varphi \in [0, \pi)$, and the computational domain is $\Omega = [-20, 20] \times [-20, 20] \times [-36, 36]$. Figures 7 and 8 show the evolution of $\phi(x, y, z, t)$ with $g = 0$ and 1, respectively. The results in Figures 7 and 8 suggest that pattern formation on a surface having narrow and sharp tips can be simulated by using our method. Figure 9 shows the energy decay with $g = 0$ and 1.

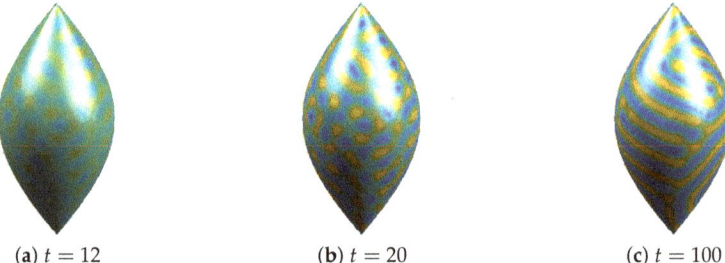

Figure 7. Evolution of $\phi(x, y, z, t)$ with $g = 0$. The yellow and blue regions indicate $\phi = 0.7059$ and -0.7593, respectively.

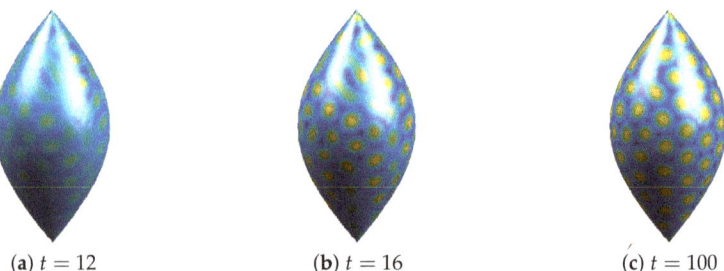

Figure 8. Evolution of $\phi(x, y, z, t)$ with $g = 1$. The yellow and blue regions indicate $\phi = 1.3842$ and -0.6224, respectively.

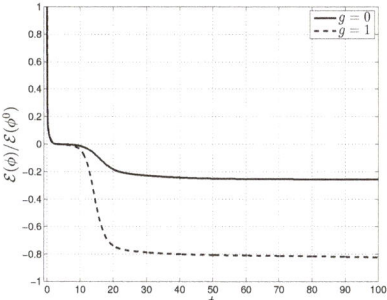

Figure 9. Evolution of $\mathcal{E}(\phi)/\mathcal{E}(\phi^0)$ on the spindle with $g = 0$ and 1.

5. Conclusions

We simulated pattern formation on surfaces numerically by solving the SH type of equation on surfaces by using the efficient linear second-order method. The method was based on the closest point and operator splitting methods and thus was easy to implement and linear. We confirmed that the proposed method gives the desired order of accuracy in time and observed that pattern formation on surfaces is affected by the value of g.

Funding: This work was supported by the Basic Science Research Program through the National Research Foundation of Korea (NRF) funded by the Ministry of Education (2019R1C1C1011112).

Acknowledgments: The corresponding author thanks the reviewers for the constructive and helpful comments on the revision of this article.

Conflicts of Interest: The author declares no conflict of interest.

References

1. Swift, J.; Hohenberg, P.C. Hydrodynamic fluctuations at the convective instability. *Phys. Rev. A* **1977**, *15*, 319–328. [CrossRef]
2. Hohenberg, P.C.; Swift, J.B. Effects of additive noise at the onset of Rayleigh–Bénard convection. *Phys. Rev. A* **1992**, *46*, 4773–4785. [CrossRef] [PubMed]
3. Cross, M.C.; Hohenberg, P.C. Pattern formation outside of equilibrium. *Rev. Mod. Phys.* **1993**, *65*, 851–1112. [CrossRef]
4. Rosa, R.R.; Pontes, J.; Christov, C.I.; Ramos, F.M.; Rodrigues Neto, C.; Rempel, E.L.; Walgraef, D. Gradient pattern analysis of Swift–Hohenberg dynamics: Phase disorder characterization. *Phys. A* **2000**, *283*, 156–159. [CrossRef]
5. Hutt, A.; Atay, F.M. Analysis of nonlocal neural fields for both general and gamma-distributed connectivities. *Phys. D* **2005**, *203*, 30–54. [CrossRef]
6. Cheng, M.; Warren, J.A. An efficient algorithm for solving the phase field crystal model. *J. Comput. Phys.* **2008**, *227*, 6241–6248. [CrossRef]
7. Wise, S.M.; Wang, C.; Lowengrub, J.S. An energy-stable and convergent finite-difference scheme for the phase field crystal equation. *SIAM J. Numer. Anal.* **2009**, *47*, 2269–2288. [CrossRef]
8. Gomez, H.; Nogueira, X. A new space–time discretization for the Swift–Hohenberg equation that strictly respects the Lyapunov functional. *Commun. Nonlinear Sci. Numer. Simul.* **2012**, *17*, 4930–4946. [CrossRef]
9. Elsey, M.; Wirth, B. A simple and efficient scheme for phase field crystal simulation. *ESAIM: Math. Model. Numer. Anal.* **2013**, *47*, 1413–1432. [CrossRef]
10. Sarmiento, A.F.; Espath, L.F.R.; Vignal, P.; Dalcin, L.; Parsani, M.; Calo, V.M. An energy-stable generalized-α method for the Swift–Hohenberg equation. *J. Comput. Appl. Math.* **2018**, *344*, 836–851. [CrossRef]
11. Lee, H.G. An energy stable method for the Swift–Hohenberg equation with quadratic–cubic nonlinearity. *Comput. Methods Appl. Mech. Eng.* **2019**, *343*, 40–51. [CrossRef]
12. Matthews, P.C. Pattern formation on a sphere. *Phys. Rev. E* **2003**, *67*, 036206. [CrossRef] [PubMed]

13. Sigrist, R.; Matthews, P. Symmetric spiral patterns on spheres. *SIAM J. Appl. Dyn. Syst.* **2011**, *10*, 1177–1211. [CrossRef]
14. Ruuth, S.J.; Merriman, B. A simple embedding method for solving partial differential equations on surfaces. *J. Comput. Phys.* **2008**, *227*, 1943–1961. [CrossRef]
15. Choi, Y.; Jeong, D.; Lee, S.; Yoo, M.; Kim, J. Motion by mean curvature of curves on surfaces using the Allen–Cahn equation. *Int. J. Eng. Sci.* **2015**, *97*, 126–132. [CrossRef]
16. Lee, H.G.; Kim, J. A simple and efficient finite difference method for the phase-field crystal equation on curved surfaces. *Comput. Methods Appl. Mech. Eng.* **2016**, *307*, 32–43. [CrossRef]
17. Pak, D.; Han, C.; Hong, W.-T. Iterative speedup by utilizing symmetric data in pricing options with two risky assets. *Symmetry* **2017**, *9*, 12. [CrossRef]
18. Zong, C.; Tang, Y.; Cho, Y.J. Convergence analysis of an inexact three-operator splitting algorithm. *Symmetry* **2018**, *10*, 563. [CrossRef]
19. Bertalmío, M.; Cheng, L.-T.; Osher, S.; Sapiro, G. Variational problems and partial differential equations on implicit surfaces. *J. Comput. Phys.* **2001**, *174*, 759–780. [CrossRef]
20. Greer, J.B. An improvement of a recent Eulerian method for solving PDEs on general geometries. *J. Sci. Comput.* **2006**, *29*, 321–352. [CrossRef]
21. Strang, G. On the construction and comparison of difference schemes. *SIAM J. Numer. Anal.* **1968**, *5*, 506–517. [CrossRef]

© 2019 by the author. Licensee MDPI, Basel, Switzerland. This article is an open access article distributed under the terms and conditions of the Creative Commons Attribution (CC BY) license (http://creativecommons.org/licenses/by/4.0/).

Article

A Numerical Solution of Fredholm Integral Equations of the Second Kind Based on Tight Framelets Generated by the Oblique Extension Principle

Mutaz Mohammad

Department of Mathematics and Statistics, College of Natural and Health Sciences, Zayed University, Abu Dhabi 144543, UAE; Mutaz.Mohammad@zu.ac.ae; Tel.: +971-2-599-3496

Received: 22 May 2019; Accepted: 27 June 2019; Published: 2 July 2019

Abstract: In this paper, we present a new computational method for solving linear Fredholm integral equations of the second kind, which is based on the use of B-spline quasi-affine tight framelet systems generated by the unitary and oblique extension principles. We convert the integral equation to a system of linear equations. We provide an example of the construction of quasi-affine tight framelet systems. We also give some numerical evidence to illustrate our method. The numerical results confirm that the method is efficient, very effective and accurate.

Keywords: Fredholm integral equations; multiresolution analysis; unitary extension principle; oblique extension principle; B-splines; wavelets; tight framelets

1. Introduction

Integral equations describe many different events in science and engineering fields. They are used as mathematical models for many physical situations. Therefore, the study of integral equations and methods for solving them are very useful in application. The aim of this paper is to present a numerical method by using tight framelets for approximating the solution of a linear Fredholm integral equation of the second kind given by

$$u(x) = f(x) + \lambda \int_a^b \mathcal{K}(x,t) u(t) dt, \quad -\infty < a \le x \le b < \infty.$$

Although many numerical methods use wavelet expansions to solve integral equations, other types of methods work better with redundant systems, of which framelets are the easiest to use. The redundant system offered by frames has already been put to excellent use for many applications in science and engineering. Reference [1], particularly, frames play key roles in wavelet theory, time frequency analysis for signal processing, filter bank design in electrical engineering, the theory of shift-invariant spaces, sampling theory and many other areas (see e.g., References [2–7]. The concept of frame can be traced back to Reference [8]. It is known that the frame system is a redundant system. The redundancy of frames plays an important role in approximation analysis for many classes of functions. In the orthonormal wavelet systems, there is no redundancy. Hence, with redundant tight framelet systems, we have more freedom in building better reconstruction and approximation order.

Since 1991, wavelets have been applied in a wide range of applications and methods for solving integral equations. A short survey of these articles can be found in References [9,10]. There is a number of approximate methods for numerically solving various classes of integral equations [11,12]. It is known that Fredholm integral equations may be applied to boundary value problems and partial differential equations in practice. Also, there is a difficulty to find the analytic solution of Fredholm integral equations. Here, we use a new and efficient method that generalizes the Galerkin-wavelet method used in the literature. We will call it the Galerkin-framelet method.

The paper is organized as follows. Section 2 is devoted to providing some preliminary background of frames, some notations and its function representation. Section 3 provides some fundamentals in the construction of quasi-affine B-spline tight framelet systems using the unitary extension principle and its generalization. We then begin the presentation of solving linear Fredholm integral equations based on the Galerkin-type method, based on quasi-affine tight framelets in Section 4. We present the error analysis of the proposed method in Section 5. We further present the numerical results and some graph illustrations in Section 6. We conclude with several remarks in Section 7.

2. Preliminary Results

Frame theory is a relatively emerging area in pure as well as applied mathematics research and approximation. It has been applied in a wide range of applications in signal processing [13], image denoising [14], and computational physics and biology [15]. Interested readers should consult the references therein to get a complete picture.

The expansion of a function in general is not unique. So, we can have a redundancy for a given representation. This happens, for instance, in the expansion using tight frames. Frames were introduced in 1952 by Duffin and Schaeffer [8]. They used frames as a tool in their paper to study a certain class of non-harmonic Fourier series. Thirty years later, Young introduced a beautiful development for abstract frames and presented their applications to non-harmonic Fourier series [16]. Daubechies et al. constructed frames for $L^2(\mathbb{R})$ based on dilations and translation of functions [17]. These papers and others spurred a dramatic development of wavelet and framelet theory in the following years.

The space $L^2(\mathbb{R})$ is the set of all functions $f(x)$ such that

$$\|f\|_{L^2(\mathbb{R})} = \left(\int_{\mathbb{R}} |f(x)|^2 \right)^{1/2} < \infty.$$

Definition 1. *A sequence $\{f_k\}_{k=1}^{\infty}$ of elements in $L^2(\mathbb{R})$ is a frame for $L^2(\mathbb{R})$ if there exist constants $A, B > 0$ such that*

$$A\|f\|^2 \leq \sum_{k=1}^{\infty} |\langle f, f_k \rangle|^2 \leq B\|f\|^2, \forall f \in L^2(\mathbb{R}).$$

A frame is called tight if $A = B$.

Let $\ell_2(\mathbb{Z})$ be the set of all sequences of the form $h[k]$ defined on $\mathbb{Z}$, satisfying

$$\Big(\sum_{k=-\infty}^{\infty} |h[k]|^2 \Big)^{1/2} < \infty.$$

The Fourier transform of a function $f \in L^2(\mathbb{R})$ is defined by

$$\widehat{f}(\xi) = \int_{\mathbb{R}} f(t) \, \mathbf{e}^{-i\xi t} dt, \; \xi \in \mathbb{R},$$

and its inverse is

$$f(x) = \frac{1}{2\pi} \int_{\mathbb{R}} \widehat{f}(\xi) \, \mathbf{e}^{i\xi x} d\xi, \; x \in \mathbb{R}.$$

Similarly, we can define the Fourier series for a sequence $h \in \ell_2(\mathbb{Z})$ by

$$\widehat{h}(\xi) = \sum_{k \in \mathbb{Z}} h[k] \, \mathbf{e}^{-i\xi k}.$$

Definition 2. *A compactly supported function $\phi \in L^2(\mathbb{R})$ is said to be refinable if*

$$\phi(x) = 2 \sum_{k \in \mathbb{Z}} h_0[k] \phi(2x - k), \qquad (1)$$

for some finite supported sequence $h_0[k] \in \ell_2(\mathbb{Z})$. The sequence h_0 is called the *low pass filter* of ϕ.

A wavelet is said to have a vanishing moment of order m if

$$\int_{-\infty}^{\infty} x^p \psi(x) dx = 0; \quad p = 0, \ldots, m-1.$$

To formulate the matrix form and the numerical solution of a given Fredholm integral equation, we will study and use tight framelets and their constructions that are derived from the unitary extension principle (UEP) and the oblique extension principle (OEP) [18]. The UEP is a method to construct tight frame wavelet filters. We recall a result by Ron and Shen, Theorem 1, which constructs a tight frame from the UEP generated by a collection $\{\psi^\ell\}_{\ell=1}^r$. The interested reader may consult References [19–21], and other related references for more details.

Let $\Psi = \{\psi^\ell\}_{\ell=1}^r \subset L^2(\mathbb{R})$ be of the form

$$\psi^\ell = 2 \sum_{k \in \mathbb{Z}} h_\ell[k] \phi(2 \cdot -k), \tag{2}$$

where $\{h_\ell[k], k \in \mathbb{Z}\}_{\ell=1}^r$ is a finitely supported sequence and is called the *high pass filter* of the system.

Theorem 1 (Unitary Extension Principle [20]). *Let $\phi \in L^2(\mathbb{R})$ be the compactly supported refinable function with its finitely supported low pass filter h_0. Let $\{h_\ell[k], k \in \mathbb{Z}, \ell = 1, \ldots, r\}$ be a set of finitely supported sequences, then the system*

$$X(\Psi) = \left\{ \psi_{j,k}^\ell : 1 \le \ell \le r \; ; j, k \in \mathbb{Z} \right\}, \tag{3}$$

where $\psi_{j,k}^\ell = 2^{j/2} \psi^\ell(2^j x - k)$, forms a tight frame for $L^2(\mathbb{R})$ provided the equalities

$$\sum_{\ell=0}^r |\widehat{h}_\ell(\xi)|^2 = 1 \quad \text{and} \quad \sum_{\ell=0}^r \widehat{h}_\ell(\xi) \widehat{h}_\ell(\xi + \pi) = 0 \tag{4}$$

hold for all $\xi \in [-\pi, \pi]$.

For the proof see [18].

This means for any $f \in L^2(\mathbb{R})$, we have the following *tight framelets representation*,

$$f = \sum_{\ell=1}^r \sum_{j \in \mathbb{Z}} \sum_{k \in \mathbb{Z}} \langle f, \psi_{j,k}^\ell \rangle \psi_{j,k}^\ell. \tag{5}$$

The representation (5) is one of many; however, it is known as the best possible representation of f, which can be truncated by S_n, where

$$S_n f = \sum_{\ell=1}^r \sum_{j<n} \sum_{k \in \mathbb{Z}} \langle f, \psi_{j,k}^\ell \rangle \psi_{j,k}^\ell, n \in \mathbb{N}, \tag{6}$$

which are known as the quasi-projection operators [4].

Note that $S_n f \in L^2(\mathbb{R})$ and $\lim \|S_n f - f\|_2 = 0$ as $n \to \infty$ [22]. We will use this representation to find the numerical solutions of a given Fredholm integral equation using the quasi-affine tight framelets generated by some refinable functions.

3. Quasi-Affine B-Spline Tight Framelet Systems

There is an interesting family of refinable functions known as B-splines. It has an important role in applied mathematics, geometric modeling and many other areas [23,24]. An investigation of the frame set using a class of functions that called generalized B-spline and which includes the B-spline has been studied extensively in Reference [25].

In applications, the B-splines of order 2 and 4 are more popular than those of other orders. Also, it is preferred to have the B-splines to be centered at $x = 0$. Therefore, we define the centered B-splines as follows:

Definition 3 ([26]). *The B-spline B_{m+1} is defined as follows by using the convolution*

$$B_{m+1}(x) := (B_m * B_1)(x), x \in \mathbb{R},$$

where $B_1(x)$ is defined to be $\chi_{[-\frac{1}{2},\frac{1}{2})}(x)$, the characteristic function for the interval $[-\frac{1}{2},\frac{1}{2})$.

Figure 1 shows the graphs of the first few B-splines.

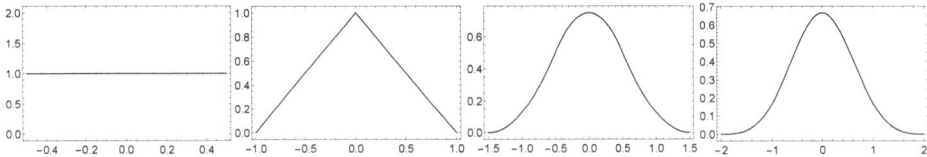

Figure 1. The B-splines B_m for $m = 1, \ldots, 4$, respectively.

One can easily show that the Fourier transform of the B-spline, B_m, of order m is given by

$$\widehat{B}_m(\xi) = e^{-i\xi d} \left(\frac{\sin(\xi/2)}{\xi/2} \right)^m \quad \text{and} \quad \widehat{h}_0^m(\xi) = e^{-i\xi d/2} \cos^m(\xi/2),$$

where $d = 0$ if m is even, and $d = 1$ if m is odd. We refer to [27] for more details.

3.1. Framelets by the UEP and Its Generalization

The UEP is a method to construct tight framelets from a given refinable function. For a given refinable function and to construct tight framelets system, the function Θ, which is non-negative, essentially bounded and continuous at the origin with $\Theta(0) = 1$, should satisfy the following conditions

$$\begin{cases} \Theta(2\xi)|\widehat{h}_0(\xi)|^2 + \sum_{\ell=1}^r |\widehat{h}_\ell(\xi)|^2 = \Theta(\xi); \\ \Theta(2\xi)\widehat{h}_0(\xi)\overline{\widehat{h}_0(\xi + \pi)} + \sum_{\ell=1}^r \widehat{h}_\ell(\xi)\overline{\widehat{h}_\ell(\xi + \pi)} = 0. \end{cases} \quad (7)$$

In applications, it is recommended to use tight framelet systems that are shift-invariant. The set of functions is said to be ρ-shift-invariant if for any $k \in \mathbb{Z}$ and $\psi \in S$, we have $\psi(\cdot - \rho k) \in S$. Hence, the quasi-affine system was introduced to convert the system $X(\Psi)$ (not shift-invariant) to a shift-invariant system. Next, we present a quasi-affine system that allows us to construct a quasi-affine tight framelet. This system is not an orthonormal basis [28].

Definition 4 ([22]). *Let Ψ be defined as in the UEP. A corresponding quasi-affine system from level J is defined as*

$$X^J(\Psi) = \left\{ \psi_{j,k}^\ell : 1 \leq \ell \leq r, j, k \in \mathbb{Z}, \right\}$$

where $\psi_{j,k}^\ell$ is given by

$$\psi_{j,k}^\ell = \begin{cases} 2^{j/2} \psi^\ell(2^j \cdot -k), & j \geq J \\ 2^j \psi^\ell(2^j (\cdot - 2^{-J} k)), & j < J \end{cases}.$$

The quasi-affine system is created by changing the basic definition of $\psi_{j,k}^{\ell}$ by sampling the tight framelet system from the level $J-1$ and below. Therefore, the system $X^J(\Psi)$ is a 2^{-J}-shift-invariant system. Note that this downward sampling in the definition will not change the approximation order as it is the same for both systems. So the error analysis will be the same. We use this definition to solve a given Fredholm integral equation that will be defined later in Section 4. In the study of our expansion method, we consider $J = 0$ where $X^0(\Psi)$ generates a quasi-affine tight framelet system for $L^2(\mathbb{R})$ but not an orthonormal basis as stated earlier. It is well known that the system $X^J(\Psi)$ is a tight framelet $L^2(\mathbb{R})$ iff $X(\Psi)$ it is a tight framelet for $L^2(\mathbb{R})$. We refer the reader to Reference [22] for full details of the relation and analysis of both affine and quasi-affine systems.

To construct quasi-affine tight framelets generated by the OEP, (see Reference [20]), we should choose Θ as a suitable approximation at $\xi = 0$ to the fraction $1/|\widehat{\phi}|$. As an example for the B-spline of order m, we should choose Θ to be a 2π-periodic function that approximates the reciprocal of

$$\left|\frac{\sin(\xi/2)}{\xi/2}\right|^{2m}$$

at $\xi = 0$.

If $\widehat{h}_0$ is the low pass filter of a refinable function, then based on the OEP and to construct quasi-affine tight framelets, it is assumed [22] that

$$H(\xi) = \Theta(\xi) - \Theta(2\xi)|\widehat{h}_0(\xi)|^2 - \Theta(2\xi)|\widehat{h}_0(\xi+\pi)|^2 \geq 0. \qquad (8)$$

This condition will help to find high pass filters easily. Let $|h|^2 = H$ and $|\theta|^2 = \Theta$. Here the square root is obtained by the Féjer-Riesz lemma [2]. Choose c_2, c_3 to be 2π-periodic trigonometric polynomials such that

$$|c_2(\xi)|^2 + |c_3(\xi)|^2 = 1, \quad c_2(\xi)\overline{c_2(\xi+\pi)} + c_3(\xi)\overline{c_3(\xi+\pi)} = 0.$$

Then, we can find three high pass filters, namely

$$\widehat{h}_1(\xi) = e^{i\xi}\theta(2\xi)\overline{\widehat{h}_0(\xi+\pi)}, \quad \widehat{h}_2(\xi) = c_2(\xi)h(\xi), \quad \widehat{h}_3(\xi) = c_3(\xi)h(\xi),$$

with a standard choice of $c_2(\xi) = (1/\sqrt{2})$, and $c_3(\xi) = (1/\sqrt{2})e^{i\xi}$.

If we consider the UEP rather than the OEP in the construction above, that is, $\Theta = 1$, then we will use the assumption that

$$|\widehat{h}_0(\xi)|^2 + |\widehat{h}_0(\xi+\pi)|^2 \leq 1.$$

Define the high pass filters as

$$\widehat{h}_1(\xi) = e^{i\xi}\overline{\widehat{h}_0(\pi+\xi)}, \quad \widehat{h}_2(\xi) = (\sqrt{2}/2)h(\xi), \quad \widehat{h}_3(\xi) = e^{i\xi}\widehat{h}_2(\xi).$$

The number of generators can be reduced from three to two with the new fundamental function $1 - H$, where

$$\widehat{h}_1(\xi) = e^{i\xi}\theta(2\xi)\overline{\widehat{h}_0(\pi+\xi)}, \quad \widehat{h}_2(\xi) = h_0(\xi)h(2\xi). \qquad (9)$$

However, this usually will affect the generators system by having less symmetry of the generators or longer filters.

3.2. Examples of Quasi-Affine B-Spline Tight Framelets

Next, we give some examples of quasi-affine B-spline tight framelets of $L^2(\mathbb{R})$ constructed via the UEP (OEP).

Example 1 (Quasi-affine HAAR framelet (HAAR framelet)). Let $h_0 = [\frac{1}{2}, \frac{1}{2}]$ be the low pass filter of $B_1(x)$. By Equations (4), we have $h_1[k] = [\frac{1}{2}, -\frac{1}{2}]$. Then, the system $X^0(\psi_1)$ forms a quasi-affine tight framelet system for $L^2(\mathbb{R})$.

Example 2 (B_2-UEP). Let $h_0 = [\frac{1}{4}, \frac{1}{2}, \frac{1}{4}]$ be the low pass filter of $B_2(x)$. We used Mathematica to obtain the high pass filters h_1 and h_2, namely, $h_1 = \pm[-\frac{1}{4}, \frac{1}{2}, -\frac{1}{4}]$; $h_2 = \pm[\frac{-\sqrt{2}}{4}, 0, \frac{\sqrt{2}}{4}]$. Considering Theorem 1, we obtain that the system $X^0(\{\psi_\ell\}_{\ell=1}^2)$ forms a quasi-affine tight framelet system for $L^2(\mathbb{R})$. B_2 and its quasi-affine tight framelet generators ψ_1, ψ_2 are illustrated in Figure 2.

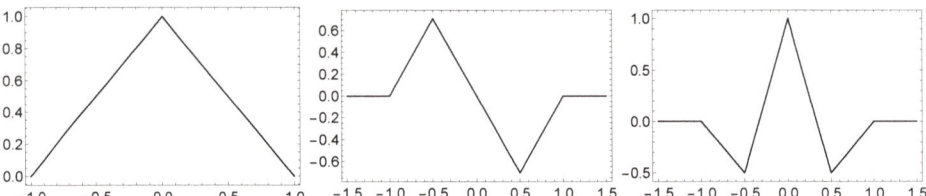

Figure 2. Piecewise linear B-spline, $B_2(x)$, and the corresponding tight framelets.

Example 3 (B_4-UEP). Let $h_0 = [\frac{1}{16}, \frac{1}{4}, \frac{3}{8}, \frac{1}{4}, \frac{1}{16}]$ be the low pass filter of $B_4(x)$. Define

$$\begin{cases} h_1 = [\frac{1}{16}, -\frac{1}{4}, \frac{3}{8}, -\frac{1}{4}, \frac{1}{16}], \; h_2 = [\frac{1}{8}, -\frac{1}{4}, 0, \frac{1}{4}, -\frac{1}{8}], \\ h_3 = [\frac{1}{8}\sqrt{\frac{3}{2}}, 0, -\frac{1}{4}\sqrt{\frac{3}{2}}, 0, \frac{1}{8}\sqrt{\frac{3}{2}}], \; h_4 = [-\frac{1}{8}, -\frac{1}{4}, 0, \frac{1}{4}, \frac{1}{8}]. \end{cases}$$

Then, h_0, h_1 and h_2 satisfy Equation (4). Hence, the system $X^0(\Psi)$ is a quasi-affine tight framelet system for $L^2(\mathbb{R})$. The cubic quasi-affine tight framelets functions, ψ_1, ψ_2, ψ_3, and ψ_4, are depicted in Figure 3.

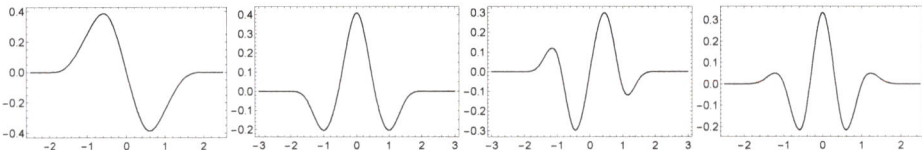

Figure 3. The corresponding tight framelets generated by the cubic B-spline.

Let us illustrate the discussion by providing some examples of quasi-affine tight framelets generated using the OEP.

Example 4 (B_2-OEP). Consider the hat function, the linear B-spline B_2, and $\Theta(\xi) = 4/3 - e^{-i\xi}/6 - e^{i\xi}/6$. Define $\Psi = \{\psi_1, \psi_2\}$, where

$$\widehat{\psi_1}(\xi) = \frac{-1}{\xi^2}\left(1 - e^{-i\xi/2}\right)^4 \text{ and } \widehat{\psi_2}(\xi) = \frac{1}{\sqrt{6}\,\xi^2}\left(-1 + e^{-i\xi/2}\right)^4\left(1 + 4e^{-i\xi/2} + e^{-i\xi}\right).$$

where θ and h in Equation (9) are obtained by using the spectral factorization theorem in [2]. Note that in time domain, we have

$$\psi_1(x) = |1 - 2x| + |3 - 2x| - \frac{1}{2}|-2 + x| - 3|-1 + x| - \frac{1}{2}|x|, \text{ and}$$

$$\psi_2(x) = \frac{1}{\sqrt{6}}\left(-4|3 - 2x| + \frac{1}{2}|-3 + x| - \frac{9}{2}|-2 + x| + \frac{1}{2}|-1 + x| + \frac{1}{2}|x|\right),$$

Then, the system $X(\Psi)$ generated using Equation (7) forms a quasi-affine tight framelet system for $L^2(\mathbb{R})$. B_2, and its quasi-affine tight framelets, ψ_1, ψ_2, are given in Figure 4.

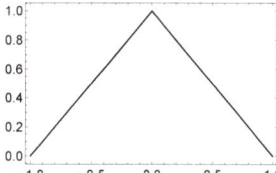 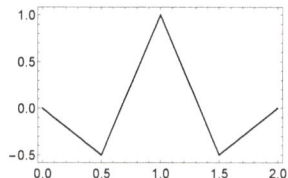 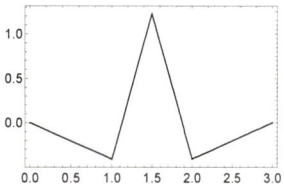

Figure 4. Piecewise linear B-spline, $B_2(x)$, and the corresponding quasi-affine tight framelets generated by the oblique extension principle (OEP).

Example 5 (B_4-OEP). *Consider $B_4(x)$, the cubic B-spline, and take the periodic function Θ defined by*

$$\Theta(\xi) = 2.59471 - 0.98631 e^{-i\xi} - 0.98631 e^{i\xi} + 0.209524 e^{-2i\xi} + 0.209524 e^{2i\xi} - 0.0205688 e^{-3i\xi} - 0.0205688 e^{3i\xi}.$$

Now, define Ψ, where $r = 3$, as follows

$$\psi_1(x) = 0.252839\,|3 - 2x|^3 + 0.707948\,|5 - 2x|^3 + 0.252839\,|7 - 2x|^3 + 0.0126419\,|-5 + x|^3 -$$
$$0.442468\,|-4 + x|^3 - 4.42468\,|-3 + x|^3 - 4.42468\,|-2 + x|^3 - 0.442468\,|1 + x|^3 +$$
$$0.0126419\,|x|^3,$$

$$\psi_2(x) = -0.07207963\,|3 - 2x|^3 - 1.21209103\,|5 - 2x|^3 - 1.21209103\,|7 - 2x|^3 - 0.07207963\,|9 - 2x|^3 +$$
$$0.01105521\,|-6 + x|^3 - 0.08269335\,|-5 + x|^3 + 3.95289189\,|-4+|^3 + 12.78422305\,|-3 + x|^3 +$$
$$3.95289189\,|-2 + x|^3 - 0.08269335\,|-1 + x|^3 + 0.01105521\,|x|^3,$$

$$\psi_3(x) = 0.0666740\,|5 - 2x|^3 + 0.798605\,|7 - 2x|^3 + 0.066674\,|9 - 2x|^3 - 2.12302 * 10^{-17}\,|11 - 2x|^3 +$$
$$0.0119515\,|-7 + x|^3 - 0.113791\,|-6 + x|^3 + 0.468945\,|-5 + x|^3 - 4.09492\,|-4 + x|^3 -$$
$$4.09492\,|-3 + x|^3 + 0.468945\,|-2 + x|^3 - 0.113791\,|-1 + x|^3 + 0.0119515\,|x|^3.$$

Then, the system $X^0(\Psi)$ satisfy Equation (7) and then forms a quasi-affine tight framelet system for $L^2(\mathbb{R})$. The cubic B-spline, B_4, and its quasi-affine tight framelet generators ψ_1, ψ_2, ψ_3 are given in Figure 5.

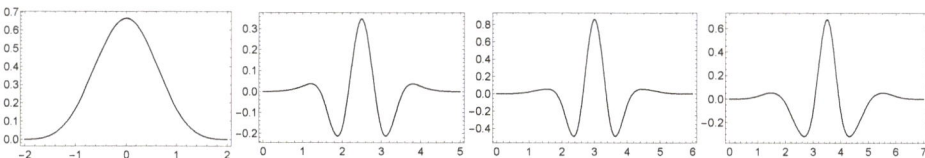

Figure 5. The B-spline $B_4(x)$ and the corresponding tight framelets generated by the OEP.

The function H in Equation (8) for the B-splines of order 2 and 4 are illustrated in Figure 6. Note that for B_2, we have

$$H(\xi) = \frac{1}{48}\left(64 - 16\,\Re(e^{-i\xi}) - 4\left(\left|1 + e^{-i\xi}\right|^4 + \left|1 - e^{i\xi}\right|^4\right) + \Re(e^{-2i\xi})\left(\left|1 + e^{-i\xi}\right|^4 + \left|1 - e^{i\xi}\right|^4\right)\right).$$

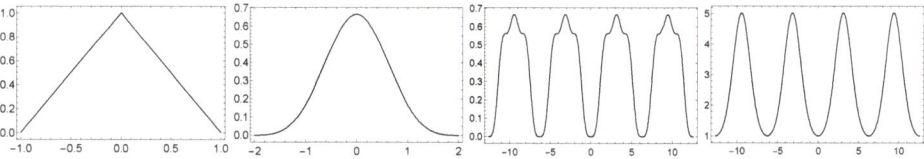

Figure 6. Illustration of B_2 and B_4, with its corresponding positive function H, respectively.

4. Solving Fredholm Integral Equation via Tight Framelets

Many methods have been presented to find exact and approximate solutions of different integral equations. In this work, we introduce a new method for solving the above-mentioned class of equations. We use quasi-affine tight framelets systems generated by the UEP and OEP for solving some types of integral equations. Consider the second-kind linear Fredholm integral equation of the form:

$$u(x) = f(x) + \lambda \int_a^b \mathcal{K}(x,t)u(t)dt, \quad -\infty < a \leq x \leq b < \infty, \tag{10}$$

where λ is a real number, f and $\mathcal{K}$ are given functions and u is an unknown function to be determined. $\mathcal{K}$ is called the kernel of the integral Equation (10). A function $u(x)$ defined over $[a,b]$ can be expressed by quasi-affine tight framelets as Equation (5). To find an approximate solution u_n of (10), we will truncate the quasi-affine framelet representation of u as in Equation (6). Then,

$$u(x) \approx u_n(x) = \sum_{\ell=1}^{r} \sum_{j<n} \sum_{k\in\mathbb{Z}} c_{j,k}^{\ell} \psi_{j,k}^{\ell}(x), \tag{11}$$

where

$$c_{j,k}^{\ell} = \int_{\mathbb{R}} u_n(x) \psi_{j,k}^{\ell}(x) dx.$$

Substituting (11) into (10) yields

$$\sum_{\ell=1}^{r} \sum_{j<n} \sum_{k\in\mathbb{Z}} c_{j,k}^{\ell} \psi_{j,k}^{\ell}(x) = f(x) + \lambda \sum_{\ell=1}^{r} \sum_{j<n} \sum_{k\in\mathbb{Z}} c_{j,k}^{\ell} \int_a^b \mathcal{K}(x,t) \psi_{j,k}^{\ell}(t) dt \tag{12}$$

Multiply Equation (12) by $\sum_{s=1}^{r} \psi_{p,q}^{s}(x)$ and integrate both sides from a to b. This can be a generalization of Galerkin method used in Reference [29,30]. Then, with a few algebra, Equation (12) can be simplified to a system of linear equations with the unknown coefficients $c_{j,k}^{\ell}$ (to be determined) given by

$$\sum_{s,\ell=1}^{r} \sum_{j<n} \sum_{k\in\mathbb{Z}} c_{j,k}^{\ell} m_{j,k,p,q}^{\ell,s} = g_{p,q}, \quad p,q \in \mathbb{Z}, \tag{13}$$

where

$$m_{j,k,p,q}^{\ell,s} = \int_a^b \psi_{j,k}^{\ell}(x) \psi_{p,q}^{s}(x) dx - \lambda \int_a^b \int_a^b \mathcal{K}(x,t) \psi_{j,k}^{\ell}(t) \psi_{p,q}^{s}(x) dxdt, \quad p,q \in \mathbb{Z}$$

and

$$g_{p,q} = \sum_{s=1}^{r} \int_a^b f(x) \psi_{p,q}^{s}(x) \, dx, \quad p,q \in \mathbb{Z}. \tag{14}$$

Note that, evaluating the values in Equation (14) and by considering the Haar framelet system, we are able to determine the values of j, k for which the representation in Equation (11) is accurate. This is done by avoiding the inner products that have zero values. If interval is $[0,1]$ and $j = -n, \ldots, n$, then $k = -2^n, \ldots, 2^n - 1$. Thus, we have a linear system of order $2^{n+1}(2n+1)$ to be solved. This system can be reduced to a smaller order by ignoring those zero inner products and again that depends on the framelet's support and the function domain being handled. Now the unknown coefficients are determined by solving the resulting system of Equation (13), and then we get the approximate solution u_n in Equation (11).

This can be formulated as a matrix form

$$M^T C = G,$$

where, for example, in the case of quasi-affine Haar framelet system, if

$j = -n, \ldots, a, \ldots, n;\ p = -n, \ldots, \alpha, \ldots, n;\ k = -2^n, \ldots, b, \ldots, 2^n - 1;\ q = -2^n, \ldots, \beta, \ldots, 2^n - 1;$
$\ell = 1, \ldots, x, \ldots, r;\ and\ s = 1, \ldots, d, \ldots, r,$

then, the matrix M and the column vectors C and G are given by

$$M = \begin{pmatrix} m^{1,1}_{-n,-2^n,-n,-2^n} & \cdots & m^{x,d}_{a,b,-n,-2^n} & \cdots & m^{r,r}_{n,2^n-1,-n,-2^n} \\ \vdots & \ddots & \vdots & \ddots & \vdots \\ m^{1,1}_{-n,-2^n,\alpha,\beta} & \vdots & m^{x,d}_{a,b,\alpha,\beta} & \vdots & m^{r,r}_{n,2^n-1,\alpha,\beta} \\ \vdots & \ddots & \vdots & \ddots & \vdots \\ m^{1,1}_{-n,-2^n,n,2^n-1} & \cdots & m^{x,d}_{a,b,n,2^n-1} & \cdots & m^{r,r}_{n,2^n-1,n,2^n-1} \end{pmatrix}$$

$$C = [c^1_{-n,-2^n}, \ldots, c^x_{a,b}, \ldots, c^r_{n,2^n-1}],$$
$$G = [g_{-n,-2^n}, \ldots, g_{a,b}, \ldots, g_{n,2^n-1}].$$

such that C and G are $2^{n+1}(2n+1) \times 1$ column vectors, and M is an $2^{n+1}(2n+1) \times 2^{n+1}(2n+1)$ matrix. The absolute error for this formulation is defined by

$$e_n = \|u(x) - u_n(x)\|_2, \quad x \in [a,b].$$

The error function is defined to be

$$E_n(x) = |u(x) - u_n(x)|, \quad x \in [a,b].$$

5. Error Analysis

In this section, we get an upper bound for the error of our method. Let ϕ be as in Equation (1) and $W_2^m(\mathbb{R})$ is the Sobolev space consists of all square integrable functions f such that $\{f^{(k)}\}_{k=0}^m \in L^2(\mathbb{R})$. Then, $X^0(\Psi)$ provides approximation order m, if

$$\|f - S_n f\|_2 \le C 2^{-nm} \|f^{(m)}\|_2, \quad \forall f \in W_2^m(\mathbb{R}), n \in \mathbb{N}.$$

The approximation order of the truncated function S_n was studied in References [20,31]. It is well known in the literature that the vanishing moments of the framelets can be determined by its low and high pass filters $\hat{h}_\ell, \ell = 0, \ldots, r$. Also, if the quasi-affine framelet system has vanishing moments of order say m_1 and the low pass filter of the system satisfy the following,

$$1 - |\hat{h}_0(\xi)|^2 = \mathcal{O}(|\cdot|^{2m}),$$

at the origin, then the approximation order of $X^0(\Psi)$ is equal to $\min\{m_1, m\}$. Therefore, as the OEP increases the vanishing moments of the quasi-affine framelet system, the accuracy order of the truncated framelet representation, will increase as well.

As mentioned earlier, integral equations describe many different events in applications such as image processing and data reconstructions, for which the regularity of the function f is low and does not meet the required order of smoothness. This makes the determination of the approximation order difficult from the functional analysis side. Instead, it is assumed that the solution function to satisfy a decay condition with a wavelet characterization of Besov space $B_{2,2}^s$. We refer the reader to Reference [32] for more details. Hence, we impose the following decay condition such that

$$N_f = \sum_{\ell=1}^{r} \sum_{j \geq 0} \sum_{k \in \mathbb{Z}} 2^{sj} \left| \langle f, \psi_{j,k}^\ell \rangle \right| < \infty, \tag{15}$$

where $s \geq -1$.

Theorem 2. *Let $X^0(\Psi)$ be a quasi-tight framelet system generated using the OEP from the compactly supported function ϕ. Assume that f satisfies the condition (15). Then,*

$$\|f - S_n f\|_2 \leq N_f \mathcal{O}(2^{-(s+1)n}). \tag{16}$$

Proof. Using Bessel property of $X(\Psi)$, and by Equation (6), we have

$$\|f - S_n f\|_2^2 = \left\| \sum_{\ell=1}^{r} \sum_{j \geq n} \sum_{k \in \mathbb{Z}} \langle f, \psi_{j,k}^\ell \rangle \psi_{j,k}^\ell \right\|_2^2$$

$$\leq \sum_{\ell=1}^{r} \sum_{j \geq n} \sum_{k \in \mathbb{Z}} \left| \langle f, \psi_{j,k}^\ell \rangle \right|^2.$$

Note that

$$\left| \langle f, \psi_{j,k}^\ell \rangle \right| \leq \|f\|_\infty \left\| \psi_{j,k}^\ell \right\|_1 = \|f\|_\infty 2^{-j/2} \left\| \psi^\ell \right\|_1.$$

This leads to the following

$$\|f - S_n f\|_2^2 \leq \|f\|_\infty \max_\ell \left\| \psi^\ell \right\|_1 \sum_{\ell=1}^{r} \sum_{j \geq n} \sum_{k \in \mathbb{Z}} 2^{-j} \left| \langle f, \psi_{j,k}^\ell \rangle \right|$$

$$\leq \|f\|_\infty \max_\ell \left\| \psi^\ell \right\|_1 \sum_{\ell=1}^{r} \sum_{j \geq n} \sum_{k \in \mathbb{Z}} 2^{-j} \frac{2^{j(s+1)}}{2^{n(s+1)}} \left| \langle f, \psi_{j,k}^\ell \rangle \right|$$

$$\leq \|f\|_\infty \max_\ell \left\| \psi^\ell \right\|_1 2^{-(s+1)n} N_f.$$

Thus, the inequality (16) is concluded. □

6. Numerical Performance and Illustrative Examples

Based on the method presented in this paper, we solve the following examples using the quasi-affine tight framelets constructed in Section 3.2. The computations associated with these examples were obtained using Mathematica software.

Example 6. *We consider the Fredholm integral equation of 2nd kind defined by:*

$$u(x) = 1 + \int_{-1}^{1} (xt + x^2 t^2) u(t) dt.$$

The exact solution is $u(x) = 1 + \frac{10}{9} x^2$.

In Tables 1 and 2 the absolute error e_n for different values of n and the numerical values of $u_n(x)$ when $n = 2$ are computed, respectively. Using quasi-affine Haar framelet system, Figures 7 and 8 demonstrated the graphs of the exact and approximate solutions and Figure 9 demonstrated the graphs of $E_n(x)$ for different values of n. For the case of B_2-UEP, Figure 10 demonstrated the graphs of the exact and approximate solutions for different values of n.

Table 1. The errors e_n of Example 6 for five different quasi-affine tight framelet systems generated by the unitary extension principle (UEP) and OEP, for increasing n.

n	HAAR Framelet	B_2-UEP	B_4-UEP	B_2-OEP	B_4-OEP
2	6.55×10^{-2}	1.83×10^{-3}	8.89×10^{-6}	1.82×10^{-3}	8.24×10^{-6}
3	3.27×10^{-2}	4.58×10^{-4}	9.09×10^{-6}	6.78×10^{-4}	1.43×10^{-7}
4	1.64×10^{-2}	1.14×10^{-4}	1.39×10^{-7}	3.25×10^{-4}	9.46×10^{-7}
5	8.18×10^{-3}	2.86×10^{-5}	2.45×10^{-8}	1.87×10^{-5}	1.03×10^{-9}
6	4.09×10^{-3}	7.15×10^{-5}	1.33×10^{-8}	7.01×10^{-6}	5.71×10^{-10}
7	1.77×10^{-4}	9.88×10^{-6}	9.33×10^{-9}	8.79×10^{-6}	9.42×10^{-11}
8	5.92×10^{-4}	5.73×10^{-6}	7.40×10^{-10}	3.08×10^{-7}	5.08×10^{-12}
9	1.70×10^{-5}	4.03×10^{-7}	4.22×10^{-11}	4.01×10^{-8}	3.32×10^{-13}
10	8.65×10^{-5}	1.21×10^{-8}	3.52×10^{-12}	1.32×10^{-9}	2.21×10^{-13}

Table 2. Numerical results of the function u_n of Example 6 using different quasi-affine tight framelets and for a level of $n = 2$.

x	Exact	HAAR Framelet	B_2-UEP	B_4-UEP	B_2-OEP	B_4-OEP
−0.9	1.9000000	1.9723995	1.8998802	1.8999970	1.8998802	1.9000001
−0.7	1.5444444	1.5235997	1.5457151	1.5444398	1.5457151	1.5444444
−0.5	1.2777777	1.2128922	1.2748830	1.2777713	1.2748830	1.2777777
−0.3	1.1000000	1.1093230	1.1012727	1.0999958	1.1012727	1.1000000
−0.1	1.0111111	1.0057538	1.0109953	1.0111080	1.0109953	1.0111100
0.0	1.0000000	1.0001228	0.9971065	0.9999965	0.9971065	1.0000000
0.1	1.0111111	1.0057538	1.0109953	1.0111078	1.0109953	1.0111110
0.3	1.1000000	1.1093230	1.1012727	1.0999938	1.1012727	1.1000000
0.5	1.2777777	1.3509844	1.2748830	1.2777699	1.2748830	1.2777777
0.7	1.5444444	1.5235997	1.5457151	1.5444346	1.5457151	1.5444442
0.9	1.9000000	1.9723995	1.8998802	1.8999885	1.8998802	1.9000000

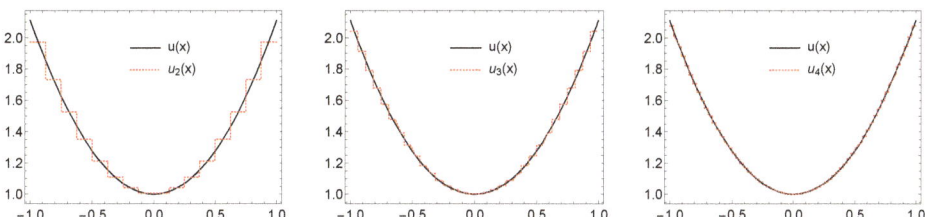

Figure 7. The graphs of u and u_n for $n = 2, 3, 4$, respectively, based on the quasi-affine HAAR framelet system of Example 6.

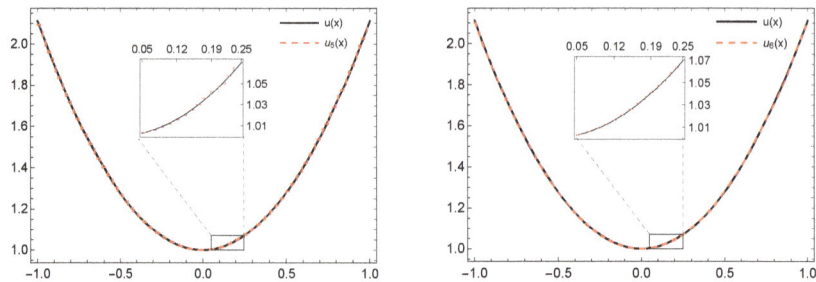

Figure 8. The graphs of u and u_n for $n = 5, 6$, respectively, based on the quasi-affine HAAR framelet system of Example 6.

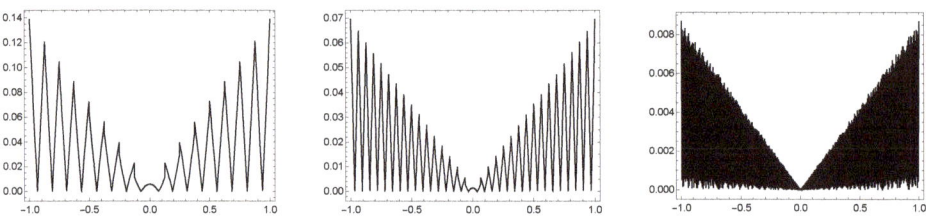

Figure 9. The graphs of $E_n(x)$ for $n = 2, 3, 6$, respectively, and based on quasi-affine HAAR framelet system of Example 6.

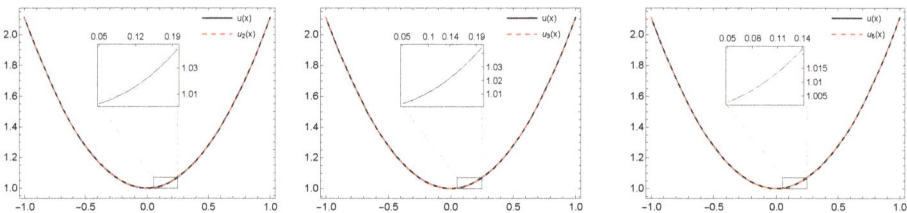

Figure 10. The graphs of u and u_n for $n = 2, 3, 5$, respectively, based on the quasi-affine B_2-UEP framelet system.

Example 7. *We consider the Fredholm integral equation of 2nd kind defined by:*

$$u(x) = e^x - \frac{e^{x+1} - 1}{x+1} + \int_0^1 e^{xt} u(t) dt.$$

The exact solution is $u(x) = e^x$.

In Tables 3 and 4 the absolute error e_n for different values of n and the numerical values of $u_n(x)$ when $n = 2$ are computed, respectively. Some illustration for the graphs of the exact and approximate solutions are depicted in Figure 11.

Table 3. The errors e_n of Example 7 for five different quasi-affine tight framelet systems generated by the UEP and OEP, for increasing n.

n	HAAR Framelet	B_2-UEP	B_4-UEP	B_2-OEP	B_4-OEP
2	6.47×10^{-2}	1.04×10^{-3}	6.07×10^{-7}	1.03×10^{-3}	4.45×10^{-7}
3	3.23×10^{-2}	4.73×10^{-4}	8.33×10^{-8}	1.08×10^{-4}	1.78×10^{-7}
4	2.68×10^{-2}	8.76×10^{-4}	3.35×10^{-8}	1.25×10^{-4}	4.35×10^{-8}
5	1.76×10^{-3}	1.56×10^{-5}	1.25×10^{-8}	9.78×10^{-5}	1.58×10^{-9}
6	2.11×10^{-4}	1.05×10^{-5}	4.33×10^{-9}	3.45×10^{-6}	6.92×10^{-11}
7	7.93×10^{-5}	6.84×10^{-6}	5.53×10^{-10}	2.39×10^{-6}	5.92×10^{-11}
8	9.02×10^{-5}	5.56×10^{-6}	5.51×10^{-11}	5.98×10^{-7}	1.82×10^{-13}
9	2.50×10^{-6}	1.98×10^{-7}	1.02×10^{-12}	5.45×10^{-8}	3.11×10^{-14}
10	4.05×10^{-7}	2.34×10^{-8}	2.42×10^{-13}	2.12×10^{-9}	4.23×10^{-15}

Table 4. Numerical results of the function u_n of Example 7 using different quasi-affine tight framelets and for a level of $n = 2$.

x	Exact	HAAR Framelet	B_2-UEP	B_4-UEP	B_2-OEP	B_4-OEP
0.0	1.00000000	1.07061012	0.99866232	0.99999601	0.99863491	0.999999958
0.1	1.10517092	1.09061851	1.10508652	1.10517150	1.10505007	1.105170028
0.2	1.22140276	1.21247432	1.22209321	1.22140205	1.22203402	1.221402071
0.3	1.34985880	1.37321902	1.35065092	1.34985878	1.35060677	1.349858253
0.4	1.49182470	1.55532117	1.49179223	1.49182450	1.49179044	1.491824418
0.5	1.64872127	1.76165283	1.64658432	1.64872193	1.64660927	1.648721644
0.6	1.82211880	1.80165032	1.82197332	1.82211894	1.82198698	1.822118840
0.7	2.01375270	2.01543677	2.01488843	2.01375233	2.01490121	2.013752248
0.8	2.22554090	2.26031276	2.22684809	2.22554035	2.22686599	2.225540081
0.9	2.45960311	2.46041532	2.45955456	2.45960353	2.45956180	2.459603410
1.0	2.71828180	2.72765982	2.71485007	2.71828155	2.71483561	2.718281710

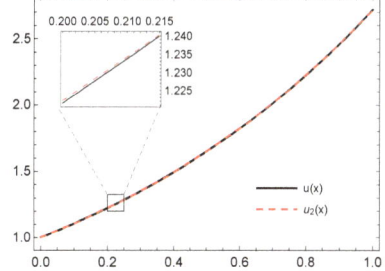
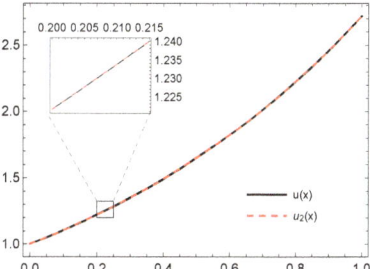

Figure 11. The graphs of u and u_n for $n = 2$, based on the quasi-affine B_2-UEP and B_4-UEP framelet systems, respectively, of Example 7.

To see the convergence of the proposed method using the quasi-affine tight framelet systems, we use the log-log scale plot. Then, it is clear that as the partial sum increases, the error between the approximated solution and exact solution approaches zero. The error history (log-log scale plot) of Examples 6 and 7 is displayed in Figure 12, where we see very accurate convergence rates too. This confirmed with respect to the theoretical predictions in the error analysis.

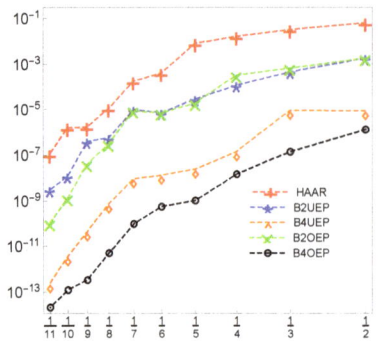 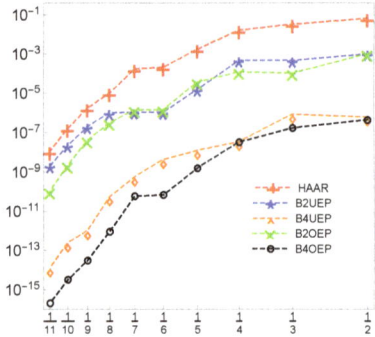

Figure 12. The convergence rate graph of Examples 6 and 7 given in the log-log scale plot, respectively.

7. Conclusions

We created a new, efficient method for solving Fredholm integral equations of the second kind. It turns out our method is efficient and has great accuracy. The proposed method shows highly accurate results and the performance of the present method is reliable, efficient and converges to the exact solution. Furthermore, the accuracy improves with increasing the partial sums and the number of vanishing moments of the *B*-splines quasi-affine tight framelets generated using the UEP and OEP.

Funding: This research was funded by Zayed University Research Fund.

Acknowledgments: The author would like to thank the anonymous reviewers for their valuable comments to improve the quality of the paper.

Conflicts of Interest: The author declare no conflict of interest.

References

1. Ionescu, M.; Okoudjou, K.A.; Rogers, L.G. Some spectral properties of pseudo-differential operators on the Sierpinski gasket. *Proc. Am. Math. Soc.* **2017**, *145*, 2183–2198. [CrossRef]
2. Daubechies, I. *Ten Lectures on Wavelets*; SIAM: Philadelphia, PA, USA, 1992.
3. Grochenig, K. *Foundations of Time-Frequency Analysis*; Birkhäuser: Boston, MA, USA, 2001.
4. Han, B. Framelets and wavelets: Algorithms, analysis, and applications. In *Applied and Numerical Harmonic Analysis*; Birkhäuser/Springer: Cham, Switzerland, 2017.
5. Mallat, S.G. *A Wavelet Tour of Signal Processing*, 2nd ed.; Elsevier: Amsterdam, The Netherlands, 1999.
6. Meyer, Y. *Wavelets and Operators*; Cambridge University Press: Cambridge, UK, 1992.
7. Meyer, Y. Oscillating Patterns in Image Processing and Nonlinear Evolution Equations: The Fifteenth Dean Jacqueline B. Lewis Memorial Lectures. American Mathematical Society. Available online: https://bookstore.ams.org/ulect-22 (accessed on 21 May 2019).
8. Duffin, R.; Schaeffer, A. A class of nonharmonic Fourier series. *Trans. Am. Math. Soc.* **1952**, *72*, 341–366. [CrossRef]
9. Adeh, Z.; Heydari, M.; Loghman, G.B. Numerical solution of Fredholm integral equations of the second kind by using integral mean value theorem. *Appl. Math. Model.* **2011**, *35*, 2374–2383.
10. Lepik, U.; Tamme, E. Application of the Haar wavelets for solution of linear integral equations. In Proceedings of the Dynamical Systems and Applications, Antalya, Turkey, 5–10 July 2005; pp. 395–407.
11. Singh, B.; Bhardwaj, A.; Alib, R. A wavelet method for solving singular integral equation of MHD. *Appl. Math. Comput.* **2009**, *214*, 271–279 [CrossRef]
12. Islam, M.S.; Shirin, A. Numerical Solutions of Fredholm Integral Equations of Second Kind Using Piecewise Bernoulli Polynomials. *Dhaka Univ. J. Sci.* **2011**, *59*, 103–107.
13. Cai, J.; Dong, B.; Shen, Z. Image restorations: A wavelet frame based model for piecewise smooth functions and beyond. *Appl. Comput. Harm. Anal.* **2016**, *41*, 94–138. [CrossRef]

14. Shen, Y.; Han, B.; Braverman, E. Adaptive frame-based color image denoising. *Appl. Comput. Harm. Anal.* **2016**, *41*, 54–74. [CrossRef]
15. Yang, J.; Zhu, G.; Tong, D.; Lu, L.; Shen, Z. B-spline tight frame based force matching method. *J. Comput. Phys.* **2018**, *362*, 208–219. [CrossRef]
16. Young, R. *An Introduction to Non-Harmonic Fourier Series*; Academic Press: New York, NY, USA, 1980.
17. Daubechies, I.; Grossmann, A.; Meyer, Y. Painless nonorthogonal expansions. *J. Math. Phys.* **1986**, *341*, 1271–1283. [CrossRef]
18. Ron, A. Factorization theorems of univariate splines on regular grids. *Isr. J. Math.* **1990**, *70*, 48–68. [CrossRef]
19. Chui, C.K.; He, W.; Stockler, J. Compactly supported tight and sibling frames with maximum vanishing moments. *Appl. Comput. Harmon. Anal.* **2002**, *341*, 224–262. [CrossRef]
20. Daubechies, I.; Han, B.; Ron, A.; Shen, Z. Framelets: MRA-based constructions of wavelet frames. *Appl. Comput. Harmon.* **2003**, *14*, 1–46. [CrossRef]
21. Ron, A.; Shen, Z. Affine systems in $L^2(\mathbb{R}^d)$ II: Dual systems. *J. Fourier Anal. Appl.* **1997**, *3*, 617–637. [CrossRef]
22. Ron, A.; Shen, Z. Affine systems in $L^2(\mathbb{R}^d)$: The analysis of the analysis operators. *J. Funct. Anal.* **1997**, *148*, 408–447. [CrossRef]
23. Mohammad, M.; Lin, E. Gibbs Phenomenon in Tight Framelet Expansions. *Commun. Nonlinear Sci. Numer. Simul.* **2018**, *55*, 84–92. [CrossRef]
24. Mohammad, M.; Lin, E. Gibbs effects using Daubechies and Coiflet tight framelet systems, Frames and Harmonic Analysis. *Contemp. Math.* **2018**, *706*, 271–282.
25. Atindehou, A.G.D.; Kouagou, Y.B.; Okoudjou, K.A. Frame sets for generalized B-splines. *arXiv* **2018**, arXiv:1804.02450.
26. He, T. Eulerian polynomials and B-splines. *J. Comput. Appl. Math.* **2012**, *236*, 3763–3773. [CrossRef]
27. De Boor, C. *A Practical Guide to Splines*; Springer: New York, NY, USA, 1978.
28. Dong, B.; Shen, Z. MRA Based Wavelet Frames and Applications. 2010. Available online: ftp://ftp.math.ucla.edu/pub/camreport/cam10-69.pdf (accessed on 21 May 2019).
29. Bhatti, M.I.; Bracken, P. Solutions of differential equations in a Bernstein polynomial basis. *J. Comput. Appl. Math.* **2007**, *205*, 272–280. [CrossRef]
30. Liang, X.Z.; Liu, M.C.; Che, X.J. Solving second kind integral equations by Galerkin methods with continuous orthogonal wavelets. *J. Comput. Appl. Math.* **2001**, *136*, 149–161. [CrossRef]
31. De Boor, C.; DeVore, R.; Ron, A. Approximation from shift-invariant subspaces of $L^2(\mathbb{R}^d)$. *Trans. Am. Math. Soc.* **1994**, *341*, 787–806.
32. Borup, L.; Gribonval, R.; Nielsen, M. Bi-framelet systems with few vanishing moments characterize Besov spaces. *Appl. Comput. Harmon. Anal.* **2004**, *17*, 3–28. [CrossRef]

© 2019 by the author. Licensee MDPI, Basel, Switzerland. This article is an open access article distributed under the terms and conditions of the Creative Commons Attribution (CC BY) license (http://creativecommons.org/licenses/by/4.0/).

Article

Qualitative Behavior of Solutions of Second Order Differential Equations

Clemente Cesarano [1,*,†] and Omar Bazighifan [2,†]

1. Section of Mathematics, International Telematic University Uninettuno, CorsoVittorio Emanuele II, 39, 00186 Roma, Italy
2. Department of Mathematics, Faculty of Science, Hadhramout University, Hadhramout 50512, Yemen; o.bazighifan@gmail.com
* Correspondence: c.cesarano@uninettunouniversity.net
† These authors contributed equally to this work.

Received: 10 May 2019; Accepted: 5 June 2019; Published: 11 June 2019

Abstract: In this work, we study the oscillation of second-order delay differential equations, by employing a refinement of the generalized Riccati substitution. We establish a new oscillation criterion. Symmetry ideas are often invisible in these studies, but they help us decide the right way to study them, and to show us the correct direction for future developments. We illustrate the results with some examples.

Keywords: second-order; nonoscillatory solutions; oscillatory solutions; delay differential equations

---

1. Introduction

This paper is concerned with oscillation of a second-order differential equation

$$[a(z) w'(z)]' + q(z) f(w(\tau(z))) = 0, \quad z \geq z_0, \tag{1}$$

where $a, \tau \in C^1([z_0, \infty), \mathbb{R}^+)$, $\tau(z) \leq z$, $\lim_{z \to \infty} \tau(z) = \infty$, $\tau'(z) \geq 0$, $q \in C([z_0, \infty), [0, \infty))$, the function f is nondecreasing and satisfies the following conditions

$$f \in C(\mathbb{R}, \mathbb{R}), \, wf(w) > 0, \, f(w)/w \geq k > 0, \, \text{for } w \neq 0, \tag{2}$$

and

$$\lim_{w \to \infty} \frac{w}{f(w)} = M < \infty, \tag{3}$$

where M is constant.

By a solution of Equation (1) we mean a function $w \in C([z_0, \infty), \mathbb{R})$, $z_w \geq z_0$, which has the property $a(z) [w'(z)] \in C^1([z_0, \infty), \mathbb{R})$, and satisfies Equation (1) on $[z_w, \infty)$. We consider only those solutions w of Equation (1) which satisfy $\sup\{|w(z)| : z \geq z_w\} > 0$. Such a solution is said to be oscillatory if it has arbitrarily large zeros and nonoscillatory otherwise.

Differential equations play an important role in many branches of mathematics, and they also often appear in other sciences. This fact leads us to more studying such equations and related boundary value problems in more detail, and a theory of solvability and (numerical) solutions for such equations are needed for distinct scientific groups, (see [1–5]).

Usually, one cannot find an exact solution for such equations, and one then needs to describe its qualitative properties in the appropriate functional spaces as well as to suggest a way of reducing the starting equation to a certain well known studied case, or to suggest some computational algorithm for the numerical solution. These studies are the intermediate points for solving equations.

The study of differential equations with deviating argument was initiated in 1918, appearing in the first quarter of the twentieth century as an area of mathematics that has received a lot of attention, (see [6–10]).

The oscillations of second order differential equations have been studied by authors and several techniques have been proposed for obtaining oscillation for these equations. For treatments on this subject, we refer the reader to the texts (see [11–15]). In what follows, we review some results that have provided the background and the motivation for the present work.

Koplatadze [16] is concerned with the oscillation of equations

$$w''(z) + p(z) w(\sigma(z)) = 0, \quad z \geq z_0,$$

and he proved it is oscillatory if

$$\limsup_{s \to \infty} \int_{\sigma(z)}^{z} \sigma(s) p(s) \, ds > 1.$$

Moaaz, et al. [17] discussed the equation

$$\left[a(z) \left(w'(z)\right)^{\beta}\right]' + p(z) f(w(\tau(z))) = 0, \quad z \geq z_0,$$

under the condition

$$\int_{z_0}^{\infty} \frac{1}{a^{\frac{1}{\beta}}(z)} dz < \infty.$$

Trench [18] used the comparison technique for the following

$$\left[a(z) w'(z)\right]' + q(z) w(\tau(z)) = 0,$$

that was compared with the oscillation of certain first order differential equation and under the condition

$$\int_{z_0}^{\infty} \frac{1}{a(z)} dz = \infty.$$

Wei in 1988 [19] discussed the equation

$$w''(z) + q(z) w(\tau(z)) = 0, \quad z \geq z_0,$$

and used the classical Riccati transformation technique.

The present authors in this paper use the generalized Riccati substitution which differs from those reported in [20–22].

This paper deals with oscillatory behavior of second order delay for Equation (1) under the condition

$$\int_{z_0}^{\infty} \frac{1}{a(s)} ds < \infty, \quad (4)$$

which would generalize and extend of the related results reported in the literature. In addition, we use a generalized Riccati substitution. Some examples are included to illustrate the importance of results obtained.

Here we mention some lemmas.

Lemma 1. *(See [23], Lemma 2.1) Let $\beta \geq 1$ be a ratio of two odd numbers, $G, H, U, V \in \mathbb{R}$. Then*

$$G^{\frac{\beta+1}{\beta}} - (G-H)^{\frac{\beta+1}{\beta}} \leq \frac{H^{\frac{1}{\beta}}}{\beta} \left[(1+\beta) G - H\right], \quad GH \geq 0,$$

and
$$Uy - Vy^{\frac{\beta+1}{\beta}} \leq \frac{\beta^\beta}{(\beta+1)^{\beta+1}} \frac{U^{\beta+1}}{V^\beta}, \quad V > 0.$$

Lemma 2. *(See [13], Lemma 1.1) Let y satisfy $y^{(i)} > 0$, $i = 0, 1, ..., n$, and $y^{(n+1)} < 0$, then*

$$\frac{y(z)}{z^n/n!} \geq \frac{y'(z)}{z^{n-1}/(n-1)!}.$$

Lemma 3. *(See [24], Lemma 1) Assume that $w(z)$ is an eventually positive solution of Equation (1). Then we have two cases*

(C_1) $a(z)w'(z) > 0$, $(a(z)w'(z))' < 0$,
(C_2) $a(z)w'(z) < 0$, $(a(z)w'(z))' < 0$, *for* $z \geq z_1 \geq z_0$.

2. Main Results

In this section, we shall establish some oscillation criteria for Equation (1). For convenience, we denote

$$B(z) := \int_z^\infty \frac{1}{a(s)} ds, \quad A(z) := kq(z)\frac{\tau^2(z)}{z^2}$$

$$\Phi(z) := \delta(z)\left[A(z) + \frac{1}{a(z)B^2(z)}\right].$$

$$\theta(z) := \frac{\delta'_+(z)}{\delta(z)} + \frac{2}{a(z)B(z)} \text{ and } \delta'_+(z) := \max\{0, \delta'(z)\}.$$

In what follows, all occurring functional inequalities are assumed to hold eventually, that is, they are satisfied for all t large enough. As usual and without loss of generality, we can deal only with eventually positive solutions of Equation (1).

Theorem 1. *Assume that Equation (3) holds and $\tau(z) \leq z$. If*

$$\limsup_{s \to \infty} \int_{\tau(z)}^z B(s)q(s)\,ds > M, \tag{5}$$

then all solutions of Equation (1) are oscillatory.

Proof. Assume, on the contrary, that $w(z)$ is an eventually positive solution of Equation (1). Since Equation (5) implies

$$\int_{z_0}^\infty B(s)q(s)\,ds < \infty,$$

thereby $w(z)$ satisfies (C_2) of Lemma 3, which yields

$$\begin{aligned} 0 &\leq w(\tau(z)) + a(\tau(z))w'(\tau(z)) \\ &\leq w(\tau(z)). \end{aligned} \tag{6}$$

Setting
$$\phi(z) = w(z) + a(z)w'(z)B(z), \tag{7}$$

we see that
$$f(\phi(\tau(z))) \leq f(w(\tau(z))). \tag{8}$$

A simple computation ensures that Equation (1) can be rewritten into the form

$$(a(z)w'(z)B(z) + w(z))' + B(z)q(z)f(w(\tau(z))) = 0, \qquad (9)$$

which in view of Equation (9) implies that $\phi(z)$ is a positive decreasing solution of the first order delay differential inequality

$$\phi'(z) + B(z)q(z)f(\phi(\tau(z))) \leq 0.$$

Integrating from $\tau(z)$ to z, we get

$$\phi(\tau(z)) \geq \int_{\tau(z)}^{z} B(s)q(s)f(\phi(\tau(s)))\,ds$$

$$\geq f(\phi(\tau(z))) \int_{\tau(z)}^{z} B(s)q(s)\,ds.$$

Thus, we obtain

$$\frac{\phi(\tau(z))}{f(\phi(\tau(z)))} \geq \int_{\tau(z)}^{z} B(s)q(s)\,ds.$$

Hence, we obtain

$$\limsup_{s \to \infty} \int_{\tau(z)}^{z} B(s)q(s)\,ds \leq \limsup_{s \to \infty} \left(\frac{\phi(\tau(z))}{f(\phi(\tau(z)))} \right),$$
$$\leq M.$$

but is a contradiction. Theorem 1 is proved. $\square$

Theorem 2. *Let Equation (4) hold and*

$$w'(z) < 0, \ (a(z)w'(z)) < 0.$$

If there exists positive function $\delta \in C^1([z_0, \infty), (0, \infty))$ such that

$$\int_{z_0}^{\infty} \left(\Phi(s) - \frac{\delta(s)a(s)(\theta(s))^2}{4} \right) ds = \infty. \qquad (10)$$

Then all solutions of Equation (1) are oscillatory.

Proof. Assume that (C_2) holds. By Lemma 2, we find

$$w(z) \geq \left(\frac{z}{2}\right) w'(z)$$

and hence

$$\frac{w'(z)}{w(z)} \leq \frac{2}{z}.$$

Integrating from $\tau(z)$ to z, we get

$$\frac{w(z)}{w(\tau(z))} \leq \frac{z^2}{\tau^2(z)}$$

and hence

$$\frac{w(\tau(z))}{w(z)} \geq \frac{\tau^2(z)}{z^2}. \qquad (11)$$

It follows from $(a(z)(w'(z))) \leq 0$, we obtain
$$w'(s) \leq \left(\frac{a(z)}{a(s)}\right) w'(z).$$

Integrating from z to z_1, we get
$$w(z_1) \leq w(z) + a(z) w'(z) \int_z^{z_1} a^{-1}(s)\, ds. \tag{12}$$

Letting $z_1 \to \infty$, we obtain
$$w(z) \geq -B(z) a(z) w'(z)$$

which implies that
$$\left(\frac{w(z)}{B(z)}\right)' \geq 0.$$

Define the function $\psi(z)$ by
$$\psi(z) := \delta(z)\left[\frac{a(z)(w'(z))}{w(z)} + \frac{1}{B(z)}\right], \tag{13}$$

then $\psi(z) > 0$ for $z \geq z_1$ and
$$\begin{aligned}
\psi'(z) &= \delta'(z)\left[\frac{a(z)(w'(z))}{w(z)} + \frac{1}{B(z)}\right] + \delta(z)\frac{(a(z) w'(z))'}{w(z)} \\
&\quad -\delta(z) a(z)\frac{(w')^2(z)}{w^2(z)} + \frac{\delta(z)}{a(z) B^2(z)} \\
&= \frac{\delta'(z)}{\delta(z)}\left(\delta(z)\left[\frac{a(z)(w'(z))}{w(z)} + \frac{1}{B(z)}\right]\right) + \delta(z)\frac{(a(z) w'(z))'}{w(z)} \\
&\quad -\delta(z) a(z)\frac{(w')^2(z)}{w^2(z)} + \frac{\delta(z)}{a(z) B^2(z)}.
\end{aligned}$$

Thus, we get
$$\begin{aligned}
\psi'(z) &= \frac{\delta'(z)}{\delta(z)}\psi(z) + \delta(z)\frac{(a(z) w'(z))'}{w(z)} \\
&\quad -\delta(z) a(z)\left[\frac{1}{a(z)}\left(\frac{a(z) w'(z)}{w(z)} + \frac{1}{B(z)}\right) - \frac{1}{a(z) B(z)}\right]^2 + \frac{\delta(z)}{a(z) B^2(z)}.
\end{aligned}$$

Using Equation (13) we obtain
$$\begin{aligned}
\psi'(z) &= \frac{\delta'(z)}{\delta(z)}\psi(z) + \delta(z)\frac{(a(z) w'(z))'}{w(z)} \\
&\quad -\delta(z) a(z)\left[\frac{\psi(z)}{\delta(z) a(z)} - \frac{1}{a(z) B(z)}\right]^2 + \frac{\delta(z)}{a(z) B^2(z)}.
\end{aligned} \tag{14}$$

Using Lemma 1 with $G = \frac{\psi(z)}{\delta(z) a(z)}$, $H = \frac{1}{a(z) B(z)}$, $\beta = 1$, we get
$$\left[\frac{\psi(z)}{\delta(z) a(z)} - \frac{1}{a(z) B(z)}\right]^2 \geq \left(\frac{\psi(z)}{\delta(z) a(z)}\right)^2 \tag{15}$$
$$-\frac{1}{a(z) B(z)}\left(\frac{2\psi(z)}{\delta(z) a(z)} - \frac{1}{a(z) B(z)}\right).$$

29

From Equations (14) and (15), we obtain

$$\psi'(z) \leq \frac{\delta'_+(z)}{\delta(z)}\psi(z) - \delta(z)kq(z)\frac{w(\tau(z))}{w(z)} - \delta(z)a(z)\left(\frac{\psi(z)}{\delta(z)a(z)}\right)^2 \qquad (16)$$
$$-\delta(z)a(z)\left[\frac{-1}{a(z)B(z)}\left(\frac{2\psi(z)}{\delta(z)a(z)} - \frac{1}{a(z)B(z)}\right)\right].$$

From Equations (11) and (16), we get

$$\psi'(z) \leq \frac{\delta'_+(z)}{\delta(z)}\psi(z) - \delta(z)A(z) - \delta(z)a(z)\left(\frac{\psi(z)}{\delta(z)a(z)}\right)^2$$
$$-\delta(z)a(z)\left[\frac{-1}{a(z)B(z)}\left(\frac{2\psi(z)}{\delta(z)a(z)} - \frac{1}{a(z)B(z)}\right)\right].$$

This implies that

$$\psi'(z) \leq \left(\frac{\delta'_+(z)}{\delta(z)} + \frac{2}{a(z)B(z)}\right)\psi(z) - \frac{1}{(\delta(z)a(z))}\psi^2(z) \qquad (17)$$
$$-\delta(z)\left[A(z) + \frac{1}{a(z)B^2(z)}\right].$$

Thus, by Equation (14) yield

$$\psi'(z) \leq -\Phi(z) + \theta(z)\psi(z) - \frac{1}{(\delta(z)a(z))}\psi^2(z). \qquad (18)$$

Applying the Lemma 1 with $U = \theta(z)$, $V = \frac{1}{(\delta(z)a(z))}$ and $y = \psi(z)$, we get

$$\psi'(z) \leq -\Phi(z) + \frac{\delta(z)a(z)(\theta(z))^2}{4}. \qquad (19)$$

Integrating from z_1 to z, we get

$$\int_{z_1}^{z}\left(\Phi(s) - \frac{\delta(s)a(s)(\theta(s))^2}{4}\right)ds \leq \psi(z_1) - \psi(z) \leq \psi(z_1),$$

which contradicts Equation (10). The proof is complete. □

Example 1. *As an illustrative example, we consider the following equation:*

$$\left(z^5 w'(z)\right)' + rzw\left(\frac{z}{2}\right) = 0, \qquad (20)$$

where $r > 0$. Let

$$a(z) = z^5, \; q(z) = rz, \; \tau(z) = \frac{z}{2}.$$

It is easy to see that all conditions of Theorem 1 are satisfied.

$$\int_z^\infty \frac{1}{a(s)}ds = \int_z^\infty \frac{1}{s^5}ds$$
$$= \frac{1}{4z^4}$$
$$< \infty,$$

and

$$\limsup_{z \to \infty} \int_{\tau(z)}^{z} B(s) q(s) \, ds$$
$$= \limsup_{z \to \infty} \int_{\frac{z}{2}}^{z} \frac{1}{4s^3} r \, ds$$
$$< \infty.$$

Hence, by Theorem 1, all solutions of Equation (20) are oscillatory.

Example 2. *We consider the equation:*

$$\left(z^2 x(z)'\right)' + x\left(\frac{z}{3}\right) = 0, \ z \geq 1. \tag{21}$$

Let

$$a(z) = z^2, \ q(z) = 1, \ \tau(z) = \frac{z}{3}.$$

If we now set $\delta(z) = 1$ and $k = 1$, then all conditions of Theorem 2 are satisfied.

$$B(z) := \int_z^\infty \frac{1}{a(s)} ds = \frac{1}{z} < \infty.$$

$$A(z) = kq(z) \left(\frac{\tau(z)}{z}\right)^2 = \frac{1}{9},$$

and

$$\int_{z_0}^\infty \left(\Phi(s) - \frac{\delta(s) a(s) (\theta(s))^{\beta+1}}{(\beta+1)^{\beta+1}}\right) ds = \infty.$$

Applying Theorem 2, we obtain that all solutions of Equation (21) are oscillatory.

Remark 1. *The results in [18] imply those in Equation (21).*

Remark 2. *The results obtained supplement and improve those in [16].*

3. Conclusions

The results of this paper are presented in a form which is essentially new and of high degree of generality. To the best of our knowledge, there are not many studies known about the oscillation of Equation (1) under the assumption of Equation (4). Our primary goal is to fill this gap by presenting simple criteria for the oscillation of all solutions of Equation (1) by using the generalized Riccati transformations which differs from those reported in [22] and using a comparison technique with first order differential equation. Further, we can consider the case of $\tau(z) \geq z$ in the future work.

Author Contributions: The authors claim to have contributed equally and significantly in this paper. All authors read and approved the final manuscript.

Funding: The authors received no direct funding for this work.

Acknowledgments: The authors thank the reviewers for for their useful comments, which led to the improvement of the content of the paper.

Conflicts of Interest: There are no competing interests between the authors.

References

1. Liang, H.; Li, Q.; Zhang, Z. New oscillatory criteria for higher-order nonlinear neutral delay differential equation. *Non. Anal.* **2008**, *69*, 1719–1731. [CrossRef]
2. Kiguradze, I.T.; Chanturiya, T.A. *Asymptotic Properties of Solutions of Nonautonomous Ordinary Differential Equations*; Kluwer Academic Publishers: Dordrecht, The Netherlands, 1993.
3. Bazighifan, O.; Elabbasy, E.M.; Moaaz, O. Oscillation of higher-order differential equations with distributed delay. *J. Inequal. Appl.* **2019**, *55*, 1–9. [CrossRef]
4. Bazighifan, O. On Oscillatory and Asymptotic Behavior of Fourth Order Nonlinear Neutral Delay Differential Equations. *Int. J. Mod. Math. Sci.* **2019**, *17*, 21–30.
5. Cesarano, C.; Foranaro, C.; Vazquez, L. Operational results in bi-orthogonal Hermite functions. *Acta Math. Univ. Comen.* **2016**, *85*, 43–68.
6. Assante, D.; Cesarano, C.; Foranaro, C.; Vazquez, L. Higher Order and Fractional Diffusive Equations. *J. Eng. Sci. Technol. Rev.* **2015**, *8*, 202–204. [CrossRef]
7. Dattoli, G.; Ricci, P.; Cesarano, C. Beyond the monomiality: The monumbrality principle. *J. Comput. Anal. Appl.* **2004**, *6*, 77–83.
8. Gyori, I.; Ladas, G. *Oscillation Theory of Delay Differential Equations with Applications*; Clarendon Press: Oxford, UK, 1991.
9. Ladde, G.; Lakshmikantham, V.; Zhang, B. *Oscillation Theory of Differential Equations with Deviating Arguments*; Marcel Dekker: New York, NY, USA, 1987.
10. Hale, J.K. Partial neutral functional differential equations. *Rev. Roum. Math. Pures Appl.* **1994**, *39*, 339–344.
11. Litsyn, E.; Stavroulakis, I.P. On the oscillation of solutions of higher order Emden–Fowler state dependent advanced differential equations. *Nonlinear Anal.* **2001**, *47*, 3877–3883. [CrossRef]
12. Cesarano, C.; Pinelas, S.; Al-Showaikh, F.; Bazighifan, O.; Bazighifan, O. Asymptotic Properties of Solutions of Fourth-Order Delay Differential Equations. *Symmetry* **2019**, *11*, 628. [CrossRef]
13. Moaaz, O.; Elabbasy, E.M.; Bazighifan, O. On the asymptotic behavior of fourth-order functional differential equations. *Adv. Differ. Equ.* **2017**, *261*, 1–13. [CrossRef]
14. Marin, M.; Ochsner, A. The effect of a dipolar structure on the Holder stability in Green-Naghdi thermoelasticity. *Cont. Mech. Thermodyn.* **2017**, *29*, 1365–1374. [CrossRef]
15. Philos, C. On the existence of nonoscillatory solutions tending to zero at ∞ for differential equations with positive delay. *Arch. Math.* **1981**, *36*, 168–178. [CrossRef]
16. Koplatadze, R. Criteria for the oscillation of Solution of differential inequalities and second order equations with retarded argument. *Tbiliss. Gos. Univ. Inst. Prikl. Mat. Trudy* **1986**, *17*, 104–121.
17. Moaaz, O.; Bazighifan, O. Oscillation criteria for second-order quasi-linear neutral functional differential equation. *Discret. Contin. Dyn. Syst. Ser. S* **2019**, *81*, 1–9.
18. Trench, W. Canonical forms and principal systems for general disconiugat equations. *Trans. Am. Math. Soc.* **1973**, *189*, 319–327. [CrossRef]
19. Wei, J.J. Oscillation of second order delay differential equation. *Ann. Differ. Equ.* **1988**, *4*, 473–478.
20. Cesarano, C.; Bazighifan, O. Oscillation of fourth-order functional differential equations with distributed delay. *Axioms* **2019**, *8*, 61. [CrossRef]
21. Fite, W. Concerning the zeros of the solutions of certain differential equations. *Trans. Am. Math. Soc.* **1918**, *19*, 341–352. [CrossRef]
22. Chatzarakis, G.E.; Dzurina, J.; Jadlovska, I. New oscillation criteria for second-order half-linear advanced differential equations. *Appl. Math. Comput.* **2019**, *347*, 404-416. [CrossRef]
23. Agarwal, R.; Zhang, C.; Li, T. Some remarks on oscillation of second order neutral differential equations. *Appl. Math. Comput.* **2016**, *274*, 178–181. [CrossRef]
24. Baculikova, B. Oscillation of second-order nonlinear noncanonical differential equations with deviating argument. *Appl. Math. Lett.* **2018**, *11*, 1–7. [CrossRef]

© 2019 by the authors. Licensee MDPI, Basel, Switzerland. This article is an open access article distributed under the terms and conditions of the Creative Commons Attribution (CC BY) license (http://creativecommons.org/licenses/by/4.0/).

Article

Approximation to Logarithmic-Cauchy Type Singular Integrals with Highly Oscillatory Kernels

SAIRA and Shuhuang Xiang *

School of Mathematics and Statistics, Central South University, Changsha 410083, China; sairahameed@csu.edu.cn
* Correspondence: xiangsh@csu.edu.cn; Tel.: +86-139-7314-3907

Received: 28 April 2019; Accepted: 22 May 2019; Published: 28 May 2019

Abstract: In this paper, a fast and accurate numerical Clenshaw-Curtis quadrature is proposed for the approximation of highly oscillatory integrals with Cauchy and logarithmic singularities, $\int_{-1}^{1} \frac{f(x)\log(x-\alpha)e^{ikx}}{x-t}dx$, $t \notin (-1,1)$, $\alpha \in [-1,1]$ for a smooth function $f(x)$. This method consists of evaluation of the modified moments by stable recurrence relation and Cauchy kernel is solved by steepest descent method that transforms the oscillatory integral into the sum of line integrals. Later theoretical analysis and high accuracy of the method is illustrated by some examples.

Keywords: Clenshaw-Curtis quadrature; steepest descent method; logarithmic singularities; Cauchy singularity; highly oscillatory integrals

1. Introduction

Boundary element method and finite element method are intensively eminent numerical approaches to evaluate partial differential equations (PDEs), which appear in variety of disciplines from engineering to astronomy and quantum mechanics [1–5]. Although these methods lead PDEs to Fredholm integral equations or Voltera integral equations, but these kind of integral equations posses integrals of oscillatory, Cauchy-singular, logarithmic singular, weak singular kernel functions. However, these classical methods are failed to approximate the integrals constitute kernel functions of highly oscillation and logarithmic singularity.

This paper aims at approximation of the integral

$$I^{\alpha}[f] = \int_{-1}^{1} \frac{f(x)\log(x-\alpha)e^{ikx}}{x-t}dx, \qquad (1)$$

where $t \in (-1,1), k \gg 1, \alpha \in [-1,1]$, $f(x)$ is relatively smooth function. For integral (1) the developed strategy for logarithmic singularity $\log(x-\alpha)$ is valid for $\alpha \in [-1,1]$. In particular, the highly oscillatory integral, $\int_{-1}^{1} f(x)e^{ikx}dx$ has been computed by many methods such as asymptotic expansion, Filon method, Levin collocation method and numerical steepest descent method [6–10]. For instant, Dominguez et al. [11] for function $f(x)$ with integrable singularities have proposed an error bound, calculated in Sobolev spaces H^m, for composite Filon-Clenshaw-Curtis quadrature. Error bound depends on the derivative of $f(x)$ and length of the interval M, for some $C_1(f)$ defined as $E_N \leq C_1(f)(\frac{1+|\log(k)|}{k^{1+\beta}})^r(\log M)^{1+\beta-r}(\frac{1}{M})^{N+1-r}$ for $\beta \in (-1,0), r \in [0,1+\beta]$.

On the other hand, one methodology for numerical evaluation of integral $\int_{-1}^{1} \frac{f(x)e^{ikx}}{x-t}dx$ is replacing $f(x)$ by different kind of polynomials [12,13]. Another technique is based on analytic continuation of the integral if the integrand $f(x)$ is analytic in the complex region [14]. As far as for $k = 0$ solution methods and properties of the solution for relative non-homogenous integrals have been discussed by using Brestain polynomials and Chebyshev polynoimals of all four kinds in [3,15].

For integral $\int_{-1}^{1} f(x) \log((x-a)^2) e^{ikx} dx$ Clenshaw-Curtise rule is applied for numerical calculation. Wherein the convergence rate is independent of k but depends on the number of nodes of quadrature rule and function $f(x)$ [16]. Furthermore, Piessense and Branders [17] established the Clenshaw-Curtis quadrature rule, relies on the recurrence relation for $\int_{-1}^{1} f(x) e^{ikx} (x+1)^\alpha \log(x+1) dx$. They replaced the nonoscillatory and nonsingular part of the integrand by Chebyshev series. Chen [18] presented the numerical approximation of the integral $I[f] = \int_{-1}^{1} \frac{f(x) e^{ikx}}{(x+1)^\alpha (x-1)^\beta \prod_{m=1}^{n}(x-\tau_m)^{\gamma_m}} dx$, with $\alpha, \beta < 1$, $a < \gamma_m < b$ and $\gamma_m \leq 1$. For analytic function $f(x)$ the integral was rewritten in the form of sum of line integrals, wherein the integrands do not oscillate and decay exponentially. Moreover, Fang [19] established the Clenshaw-Curtis quadrature for $\int_{-1}^{1} \frac{(x+1)^\alpha (x-1)^\beta f(x) \log(x+1) e^{ikx}}{x-t} dx$ for general function $f(x)$ where steepest descent method is illustrated for analytic function $f(x)$. Recently, John [20] introduced the algorithm for integral approximation of Cauchy-singular, logarithmic-singular, Hadamard type and nearly singular integrals having integrable endpoints singularities i.e., $(1-x)^\alpha (1+x)^\beta$, $(\alpha, \beta > -1)$. Composed Gauss-Jacobi quadrature consists of approximating the function $f(x)$ by Jacobi polynomials $\{P_n^{\alpha,\beta}\}_{n=0}^{N-1}$ of degree $N-1$.

However, all these proposed method are inadequate to apply directly on integral (1) in the presence of oscillation and other singularities. This work presents Clenshaw-Curtis quadrature to get recurrence relation to compute the modified moments, that takes just $O(N \log N)$ operations. The initial Cauchy singular values for recurrence relation are obtained by the steepest descent method, as it prominently renowned to evaluate highly oscillatory integrals when the integrands are analytic in sufficiently large region.

The rest of the paper is organized as follows. Section 2 delineates the quadrature algorithm for integral (1). Numerical calculation of the modified moments with recurrence relation by using some Chebyshev properties is defined. Also steepest descent method is established for Cauchy singularity where later the obtained line integrals are further approximated by generalized Gauss quadrature. Section 3 alludes some error bounds derived in terms of Clenshaw-Curtis points and the rate of oscillation k. In Section 4, numerical examples are provided to demonstrate the efficiency and accuracy of the presented method.

2. Numerical Methods

In the computation of integral $I^\alpha[f]$, the Clenshaw-Curtis quadrature approach is extensively adopted. The scheme is postulated on interpolating the function $f(x)$ at Clenshaw-Curtis points set $X_{N+1} = \{x_j = \cos \frac{j\pi}{N}\}_{j=0}^{N}$. Writing the interpolation polynomial as basis of Chebyshev series

$$f(x) \approx P_N(x) = \sum_{n=0}^{N} {}''a_n T_n(x), \qquad (2)$$

where $T_n(x)$ is the Chebyshev polynomial of first kind of degree N and double prime denotes a sum whose first and last terms are halved, the coefficients

$$a_n = \frac{2}{N} \sum_{j=0}^{N} {}''f(x_j) T_n(x_j) \qquad (3)$$

can be computed efficiently by FFT in $O(N \log N)$ operations [8,9]. This paper appertains to Clenshaw-Curtis quadrature, which depends on Hermite interpolating polynomial that allow us to get higher order accuracy

$$\tilde{P}(x_j) = f(x_j), j = 0, \cdots N; \qquad \tilde{P}(t) = f(t). \tag{4}$$

For any fixed t, we can elect felicitous N such that $t \notin \{x_j\}_{j=0}^{N}$ and rewrite Hermite interpolating polynomial of degree $N+1$ in terms of Chebyshev series

$$\tilde{P}_{N+1}(x) = \sum_{n=0}^{N+1} c_n T_n(x) \tag{5}$$

c_n can be calculated in $O(N)$ operations once if a_n are known [13,21]. Finally Clenshaw-Curtis quadrature for integral $I^\alpha[f]$ is defined as

$$I^\alpha_{N+1}[f] = \sum_{n=0}^{N+1} c_n \int_{-1}^{1} \frac{T_n(x) \log(x-\alpha) e^{ikx}}{x-t} dx$$
$$= \sum_{n=0}^{N+1} c_n D^\alpha_n(k,t) \tag{6}$$

where

$$D^\alpha_n(k,t) = \int_{-1}^{1} \frac{T_n(x) \log(x-\alpha) e^{ikx}}{x-t} dx \tag{7}$$

more specifically $D^\alpha_n(k,t)$ are called the modified moments. Efficiency of the Clenshaw-Curtis quadrature depends on the fast computation of the moments. In ensuing sub-section, we deduce the recurrence relation for $D^\alpha_n(k,t)$.

Computation of the $D^\alpha_n(k,t)$ Moments

A reputed property of Chebyshev polynomial [22]

$$T_n(x) = \frac{1}{2}(U_n(x) - U_{n-2}(x)), \tag{8}$$

leads the modified moments $D^\alpha_n(k,t) = \int_{-1}^{1} \frac{T_n(x) \log(x-\alpha) e^{ikx}}{x-t} dx$ to

$$\int_{-1}^{1} \frac{T_n(x) \log(x-\alpha) e^{ikx}}{x-t} dx = \frac{1}{2}\left(\int_{-1}^{1} \frac{U_n(x) \log(x-\alpha) e^{ikx}}{x-t} dx - \int_{-1}^{1} \frac{U_{n-2}(x) \log(x-\alpha) e^{ikx}}{x-t} dx\right)$$
$$D^\alpha_n(k,t) = \frac{1}{2}(Q^\alpha_n(k,t) - Q^\alpha_{n-2}(k,t)). \tag{9}$$

Forthcoming theorem defines the procedure to calculate the moments $Q^\alpha_n(k,t) = \int_{-1}^{1} \frac{U_n(x) \log(x-\alpha) e^{ikx}}{x-t} dx$.

Proposition 1. *The sequence $Q^\alpha_n(k,t) = \int_{-1}^{1} \frac{U_n(x) \log(x-\alpha) e^{ikx}}{x-t} dx$ satisfies the recurrence relation*

$$Q^\alpha_{n+1}(k,t) = 2Q^\alpha_n(k) + 2tQ^\alpha_n(k,t) - Q^\alpha_{n-1}(k,t), \qquad n \geq 1$$
$$Q^\alpha_1(k,t) = 2Q^\alpha_0(k) + 2tQ^\alpha_0(k,t). \tag{10}$$

where
$$Q_n^\alpha(k) = \int_{-1}^{1} U_n(x)\log(x-\alpha)e^{ikx}dx, \quad Q_0^\alpha(k) = \int_{-1}^{1}\log(x-\alpha)e^{ikx}dx. \tag{11}$$

Proof. Using Chebyshev recurrence relation
$$U_{n+1}(x) = 2xU_n(x) - U_{n-1}(x),$$

$$Q_{n+1}^\alpha(k,t) = \int_{-1}^{1}\frac{2(x-t+t)U_n(x)\log(x-\alpha)e^{ikx}}{x-t}dx - \int_{-1}^{1}\frac{U_{n-1}(x)\log(x-\alpha)e^{ikx}}{x-t}dx,$$

$$= \int_{-1}^{1} 2U_n(x)\log(x-\alpha)e^{ikx}dx + \int_{-1}^{1}\frac{2tU_n(x)\log(x-\alpha)e^{ikx}}{x-t}dx$$

$$- \int_{-1}^{1}\frac{U_{n-1}(x)\log(x-\alpha)e^{ikx}}{x-t}dx,$$

$$Q_{n+1}^\alpha(k,t) = 2Q_n^\alpha(k) + 2tQ_n^\alpha(k,t) - Q_{n-1}^\alpha(k,t).$$

□

The proof completes with the initial values $U_0(x) = 1, U_1(x) = 2x$. The starting values $Q_0^\alpha(k,t)$ and $Q_0^\alpha(k)$ of recurrence relation can be calculated by steepest descent method.

Proposition 2. *Suppose that $f(x)$ is an analytic function in the half-strip of the complex plan, $a \leq \Re(x) \leq b$ and $\Im(x) \geq 0$, and satisfies the condition for constant M and $0 \leq k_0 < k$*

$$\int_{-1}^{1}|f(x+iR)|dx \leq Me^{k_0 R},$$

then the integral (1) for $\alpha \in [-1,1]$ can be transformed into

$$\begin{aligned}I^{\pm 1}[f] &= M_1^{\pm 1} + M_2^{\pm 1} + i\pi e^{ikt}f(t)\log(t-(\pm 1)),\\ I^\alpha[f] &= N_1 + N_2 + i\pi e^{ikt}f(t)\log(t-\alpha),\end{aligned} \tag{12}$$

where

$$\begin{aligned}N_1 &= \frac{i}{k}e^{-ik}\int_{kr}^{kR}\frac{f(-1+\frac{i}{k}x)\log(-1+\frac{i}{k}x-\alpha)e^{-x}}{-1+\frac{i}{k}x-t}dx,\\ N_2 &= -\frac{i}{k}e^{ik}\int_{kr}^{kR}\frac{f(1+\frac{i}{k}x)\log(1+\frac{i}{k}x-\alpha)e^{-x}}{1+\frac{i}{k}x-t}dx,\\ M_1^{\pm 1} &= \pm\frac{i}{k}e^{\mp ik}\int_{kr}^{kR}\frac{f(\mp 1+\frac{i}{k}x)\log(\mp 2+\frac{i}{k}x)e^{-x}}{\mp 1+\frac{i}{k}x-t}dx,\\ M_2^{\pm 1} &= \mp\frac{i}{k}e^{\pm ik}\int_{kr}^{kR}\frac{f(\pm 1+\frac{i}{k}x)\log(\frac{i}{k}x)e^{-x}}{\pm 1+\frac{i}{k}x-t}dx.\end{aligned} \tag{13}$$

Proof. Following proof asserts the results for case $\alpha = 1$, and for $\alpha \in [-1, 1)$ the same technique can be used as well. Since the integrand $\frac{f(x)\log(x-1)e^{ikx}}{x-t}$ is analytic in the half strip of the complex plane, by Cauchy's Theorem, we have

$$\int_{\Gamma_1+\Gamma_2+\Gamma_3+\Gamma_4-\Gamma_5-\Gamma_6-\Gamma_7} \frac{f(x)\log(x-1)e^{ikx}}{x-t} dx = 0, \qquad (14)$$

with all the contours taken in clockwise direction (Figure 1).

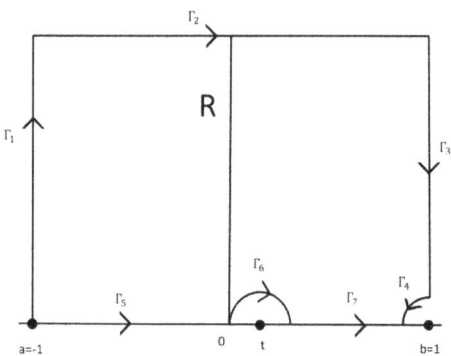

Figure 1. Illustration of integration path of $I^{+1}[f]$.

Setting $\hat{I}_i = \int_{\Gamma_i} \frac{f(x)\log(x-1)e^{ikx}}{x-t} dx, i = 1, 2, \cdots 7$, we obtain that

$$\hat{I}_1 + \hat{I}_2 + \hat{I}_3 + \hat{I}_4 = \hat{I}_5 + \hat{I}_6 + \hat{I}_7. \qquad (15)$$

$$\hat{I}_1 = \int_r^R \frac{f(-1+ip)\log(-1+ip-1)e^{ik(-1+ip)}}{-1+ip-t} i\, dp$$

$$= \frac{i}{k} e^{-ik} \int_{kr}^{kR} \frac{f(-1+\frac{i}{k}x)\log(-2+\frac{i}{k}x)e^{-x}}{-1+\frac{i}{k}x-t} dx.$$

Similarly for $\hat{I}_3$, we get

$$\hat{I}_3 = -\int_r^R \frac{f(1+ip)\log(1+ip-1)e^{ik(1+ip)}}{1+ip-t} i\, dp$$

$$= -\frac{i}{k} e^{ik} \int_{kr}^{kR} \frac{f(1+\frac{i}{k}x)\log(\frac{i}{k}x)e^{-x}}{1+\frac{i}{k}x-t} dx.$$

From the statement of the theorem, $\int_{-1}^{1} |f(x+iR)| \le M e^{w_0 R}$,

$$\hat{I}_2 = \int_{-1}^{1} \frac{f(x+iR)\log(x+iR-1)e^{ik(x+iR)}}{x+iR-t} dx$$

$$= \frac{1}{R} \int_{-1}^{1} f(x+iR)\log(x+iR-1)e^{ik(x+iR)} dx$$

$$\to 0 \quad \text{as } R \to \infty.$$

Let $x - 1 = re^{i\theta}$, then

$$\hat{I}_4 = \int_0^{\frac{\pi}{2}} \frac{f(re^{i\theta}+1)\log(re^{i\theta})e^{ik(re^{i\theta}+1)}}{1+re^{i\theta}-t} \cdot ire^{i\theta} d\theta$$

$$= ir \int_0^{\frac{\pi}{2}} \frac{f(re^{i\theta}+1)\log(re^{i\theta})e^{ik(re^{i\theta}+1)}}{1+re^{i\theta}-t} \cdot e^{i\theta} d\theta$$

$$\to 0 \quad \text{as } r \to 0.$$

In addition

$$\hat{I}_6 = \int_0^\pi \frac{f(re^{i\theta}+t)\log(t+re^{i\theta}-1)e^{ik(re^{i\theta}+t)}}{re^{i\theta}} \cdot ire^{i\theta} d\theta$$

$$r \to 0$$

$$= i\pi e^{ikt} f(t) \log(t-1).$$

Thus, we complete the proof with

$$I^{+1}[f] = \lim_{r \to 0, R \to \infty} (\hat{I}_1 + \hat{I}_2 + \hat{I}_3 + \hat{I}_4 - \hat{I}_6) \quad (16)$$

$$= M_1^{+1} + M_2^{+1} + i\pi e^{ikt} f(t) \log(t-1).$$

□

From Proposition 2.2 numerical scheme for the line integrals $M_1^{\pm 1}, M_2^{\pm 1}$ can be evaluated by generalized Gauss-Laguerre quadrature rule, using command lagpts in Chebfun [23]. Let $\{x_j^{(\beta)}, w_j^{(\beta)}\}_{j=1}^N$ be the nodes and weights of the weight function $x^\beta e^{-x}$ and let $\{x_j^{(\beta,l)}, w_j^{(\beta,l)}\}_{j=1}^N$ be the nodes and weights of the weight function $x^\beta(x-1-\ln(x))e^{-x}$. The line integrals $M_1^{\pm 1}$ and $M_2^{\pm 1}$ can be approximated by

$$M_1^{\pm 1} \approx R_{\{1,N\}}^{\pm 1} = \pm \frac{i}{k} e^{\mp ik} \sum_{j=1}^N \frac{w_j^{(\beta)} f(\mp 1 + \frac{i}{k} x_j^{(\beta)}) \log(\mp 2 + \frac{i}{k} x_j^{(\beta)})}{\mp 1 + \frac{i}{k} x_j^{(\beta)} - t} dx,$$

$$M_2^{\pm 1} \approx R_{\{2,N\}}^{\pm 1} = \mp \frac{i}{k} e^{ik} \left[\log(\frac{i}{k}) - 1 \right) \sum_{j=1}^N w_j^{(\beta)} \frac{f(\pm 1 + \frac{i}{k} x_j^{(\beta)})}{\pm 1 + \frac{i}{k} x_j^\beta - t} dx \quad (17)$$

$$+ \sum_{j=1}^N w_j^{(\beta+1)} \frac{f(\pm 1 + \frac{i}{k} x_j^{(\beta+1)})}{\pm 1 + \frac{i}{k} x_j^{(\beta+1)} - t} dx - \sum_{j=1}^N w_j^{(\beta,l)} \frac{f(\pm 1 + \frac{i}{k} x_j^{(\beta,l)})}{\pm 1 + \frac{i}{k} x_j^{(\beta,l)} - t} dx \bigg].$$

For simplicity

$$I^{\pm 1}[f] = R_{\{1,N\}}^{\pm 1} + R_{\{2,N\}}^{\pm 1} + i\pi f(t) \log(t - (\pm 1)). \quad (18)$$

By the same argument N_1 and N_2 can also be approximated with generalized Gauss-Laguerre quadrature rule. Aforementioned theorem enlightens the another interesting fact that $I^\alpha[f]$ can also be computed by it if $f(x)$ is an analytic function.

Computation of the moments $Q_n^\alpha(k)$ is derived as, by using Chebyshev property (8)

$$\frac{1}{2}(Q_n^\alpha(k) - Q_{n-2}^\alpha(k)) = D_n^\alpha(k)$$
$$= \int_{-1}^{1} (T_n(x) - T_n(\alpha)) \log(x - \alpha) e^{ikx} dx + T_n(\alpha) \int_{-1}^{1} \log(x - \alpha) e^{ikx} dx. \quad (19)$$

For $\alpha \neq \pm 1$, integrating by parts, we derive

$$\int_{-1}^{1} (T_n(x) - T_n(\alpha)) \log(x - \alpha) e^{ikx} dx = \frac{1}{ik}\left[(T_n(x) - T_n(\alpha)) \log(x - \alpha) e^{ikx} \Big|_{-1}^{1} \right.$$
$$\left. - \int_{-1}^{1} T_n'(x) \log(x - \alpha) e^{ikx} dx - \int_{-1}^{1} \frac{(T_n(x) - T_n(\alpha))}{(x - \alpha)} e^{ikx} dx \right]$$

$$= \frac{1}{ik}\left[(1 - T_n(\alpha)) \log(1 - \alpha) e^{ik} + ((-1)^{n+1} + T_n(\alpha)) \log(-1 - \alpha) e^{-ik} \right.$$
$$\left. - n \int_{-1}^{1} U_{n-1}(x) \log(x - \alpha) e^{ikx} dx - 2 \int_{-1}^{1} U_{n-1}(x) e^{ikx} dx - 2 \sum_{j=0}^{n-2} T_{n-1-j}(\alpha) \int_{-1}^{1} U_j(x) e^{ikx} dx \right]. \quad (20)$$

We deduce the following recurrence relation by inserting (20) in (19)

$$Q_n^\alpha(k) - \frac{2n}{ik} Q_{n-1}^\alpha(k) + Q_{n-2}^\alpha(k) = \delta_n^\alpha(k) \quad (21)$$

where

$$\delta_n^\alpha(k) = \frac{2}{ik}\left[(1 - T_n(\alpha)) \log(1 - \alpha) e^{ik} + ((-1)^{n+1} + T_n(\alpha)) \log(-1 - \alpha) e^{-ik} \right]$$
$$- \frac{2}{ik}\left[2 \sum_{j=0}^{n-2} T_{n-1-j}(\alpha) B_j(k) + B_{n-1}(k) \right] + 2 T_n(\alpha) Q_0^\alpha(k), \quad (22)$$

and

$$B_j(k) = \int_{-1}^{1} U_j(x) e^{ikx} dx, \quad j = 0, \cdots, n - 1. \quad (23)$$

It is worth to mention that $(B_j(k))_{j=0}^N$ can be computed in $O(N)$ operations [12]. For $\alpha = \pm 1$ we obtain the $\delta_n^{\pm 1}(k)$ as

$$\delta_n^{\pm 1}(k) = \begin{cases} 2 \log(\mp 2) e^{\mp ik} & n = odd \\ 0 & n = even. \end{cases} \quad (24)$$

Unfortunately, practical experiments demonstrate that the recurrence relation for $Q_n^\alpha(k)$ is numerically unstable in the forward direction for $n > k$, in this sense so-called Oliver's algorithm is stable and used to rewrite the recurrence relation in the tridiagonal form [24].

3. Error Analysis

Lemma 1. ([9,13,14]) Suppose $f \in C^{m+1}[-1,1]$, for a non-negative integer m with $f(t) = 0$, then

$$\left|\left(\frac{f(x)}{x-t}\right)^{(m)}\right| \leq \frac{2^{m+1}-1}{m+1}\|f^{(m+1)}\|_\infty. \tag{25}$$

Lemma 2. ([9,14]) Let $f(x)$ be a Lipschitz continuous function on $[-1,1]$ and let $P_N[f]$ be the interpolation polynomial of $f(t)$ at $N+1$ Clenshaw-Curtis points. Then it follows that

$$\lim_{N\to+\infty} \|f - P_N[f]\|_\infty = 0. \tag{26}$$

In particular, if $f(x)$ is analytic with $|f(t)| \leq M$ in an Bernstein ellipse ε_ρ with foci ± 1 and major and minor semiaxis lengths summing to $\rho > 1$, then

$$\|f - P_N[f]\|_\infty \leq \frac{4M}{\rho^N(\rho-1)}. \tag{27}$$

if $f(x)$ has an absolutely continuous $(\kappa_0 - 1)$st derivative and $f^{(\kappa_0)}$ of bounded variation V_{κ_0} on $[-1,1]$ for some $\kappa_0 \geq 1$, then for $N \geq \kappa_0 + 1$

$$\|f - P_N[f]\|_\infty \leq \frac{4V_{\kappa_0}}{\kappa_0 \pi N(N-1)\cdots(N-\kappa_0+1)}. \tag{28}$$

Lemma 3. (van der Corput Lemma [25]) Suppose that $f \in C^1[0,b]$, then for each $\beta > -1$, it follows

$$\left|\int_0^b x^\beta e^{ikx} dx\right| \leq W_1(k)\left(|f(b)| + \int_0^b |f'(x)|dx\right), \left|\int_0^b x^\beta \log(x) e^{ikx} dx\right| \leq W_2(k)\left(|f(b)| + \int_0^b |f'(x)|dx\right), \tag{29}$$

where

$$W_1(k) = \begin{cases} O(|k|^{-1-\beta}), & -1 < \beta \leq 0 \\ O(|k|^{-1}), & \beta > 0 \end{cases}, W_2(k) = \begin{cases} O(|k|^{-1-\beta}(1+|\log(k)|)), & -1 < \beta \leq 0 \\ O(|k|^{-1}), & \beta > 0 \end{cases}.$$

Moreover, for some special cases we have

Lemma 4. Suppose that $f \in C^1[0,1]$, then it follows for all k that

$$\int_0^1 x(1-x)\log(x)e^{ikx} dx = O(|k|^{-2}(1+\log|k|)), \int_0^1 x(1-x)\log(x-1)e^{ikx} dx = O(|k|^{-2}(1+\log|k|)). \tag{30}$$

Proof. For simplicity, here we prove the first identity in (3.29). Similar proof can be directly applied to the second identity in (3.29).

Since

$$\int_0^1 x(1-x)\log(x)e^{ikx} dx = \frac{1}{ik}\int_0^1 x(1-x)\log(x)de^{ikx} = -\frac{1}{ik}\int_0^1 e^{ikx}[(1-x)\log(x) - x\log(x) + (1-x)]dx,$$

it leads to the desired result by Lemma 3.3. □

Suppose that $t \notin X_{N+1}$, $f \in C^2[-1,1]$ and define

$$\phi(x) = \begin{cases} \frac{f(x)-f(t)}{(x-t)}, & x \neq t \\ f'(t), & x = t. \end{cases}$$

From Lemma 2.1, we see that $\phi \in C^1[-1,1]$ and $\|\phi'\|_\infty \leq \frac{3}{2}\|f''\|_\infty$, in addition, $g(x) = \frac{\tilde{P}_{N+1}(x)-f(t)}{x-t}$ is a polynomial of degree at most N with $g(x_j) = \phi(x_j)$ for $j = 0,1,\ldots,N$ [21]. Then the error on the Clenshaw-Curtis quadrature (6) can be estimated by

$$\begin{aligned} |E_{N+1}| = |I^\alpha[f] - I^\alpha_{N+1}[f]| &= \left|\int_{-1}^1 (\phi(x) - g(x))\log(x-\alpha)e^{ikx}dx\right| \\ &\leq \|\phi(x) - g(x)\|_\infty \left|\int_{-1}^1 \log(x-\alpha)dx\right| \\ &= O(\|\phi(x) - g(x)\|_\infty). \end{aligned}$$

Corollary 1. *Suppose that $t \notin X_{N+1}$ and f'' is bounded on $[-1,1]$, then the Clenshaw-Curtis quadrature (6) is convergent*

$$\lim_{N \to +\infty} |E_{N+1}| = \lim_{N \to +\infty} |I^\alpha[f] - I^\alpha_{N+1}[f]| = 0. \tag{31}$$

In particular, if $f(x)$ is analytic and $|f'(x)| \leq M$ in a Bernstein ellipse ε_ρ, $\rho > 1$, then the error term satisfies

$$E_{N+1} = O\left(\frac{1}{\rho^N}\right). \tag{32}$$

If $f(x)$ has an absolutely continuous $(\kappa_0 - 1)$st derivative and $f^{(\kappa_0)}$ of bounded variation V_{κ_0} on $[-1,1]$ for some $\kappa_0 \geq 1$, then for $N \geq \kappa_0 + 1$ ($\kappa_0 \geq 2$)

$$E_{N+1} = O\left(\frac{1}{N^{-\kappa_0+1}}\right). \tag{33}$$

Theorem 1. *The error bound for $I^\alpha_{N+1}[f]$ for integral $I^\alpha[f]$ can be estimated as*

$$E_{N+1} = \begin{cases} O\left(k^{-1}(1+|\log(k)|)\rho^{-N}\right), & f(x) \text{ analytic in the Bernstein ellipse } \varepsilon_\rho \\ O\left(k^{-1}(1+|\log(k)|)N^{-\kappa_0+2}\right), & f^{(\kappa_0+1)} \text{ of bounded variation} \end{cases}. \tag{34}$$

In addition, for $\alpha = \pm 1$, it follows

$$E_{N+1} = O\left(k^{-2}(1+|\log(k)|)\right) f \in C^2[-1,1]. \tag{35}$$

Proof. Since

$$\begin{aligned} E_{N+1} = \int_{-1}^1 (\phi(x) - P_N(x))\log(x-\alpha)e^{ikx}dx &= \int_{-1}^\alpha (\phi(x) - P_N(x))\log(x-\alpha)e^{ikx}dx \\ &+ \int_\alpha^1 (\phi(x) - P_N(x))\log(x-\alpha)e^{ikx}dx, \end{aligned}$$

by Lemma 3.3, it implies

$$E_{N+1} = O\left(k^{-1}(1+|\log(k)|)(\|\phi - P_N\|_\infty + \|\phi' - P'_N\|_\infty)\right),$$

which yields (3.33) together with the estimate on $\|\phi' - P'_N\|_\infty$ in [14].

The identity (3.34) follows from Lemma 3.4 due to that $\|\phi(x) - P_N(x)\| = (1+x)(1-x)h(x)$ for some $h \in C^1[-1,1]$. □

Remark 1. *From the convergence rates Corollary 3.1 and Theorem 3.1, compared with that in [19], the new scheme is of much fast convergence rate. It is also illustrated by the numerical results (see Section 4).*

4. Numerical Results

In this section, we will present several examples to illustrate the efficiency and accuracy of the proposed method. The exact values of an integral (36) are computed through Mathematica 11. Unless otherwise specifically stated, all the tested numerical examples are executed by using Matlab R2016a on a 4 GHz personal laptop with 8 GB of RAM.

Example 1. *Let us consider the integral*

$$I[f] = \int_{-1}^{1} \frac{\sin(x)\log(x-\alpha)e^{ikx}}{x-t} dx, \tag{36}$$

for $\alpha = -1$, $t = 0.3$, Table 1 shows the results for relative error compared with results of integral (30) [19] in Table 2.

Table 1. The relative error of Clenshaw-Curtis quadrature rule for integral (36).

k	N = 4	N = 7	N = 11	N = 16
20	5.642×10^{-6}	1.432×10^{-8}	1.299×10^{-13}	1.795×10^{-14}
100	1.819×10^{-7}	8.954×10^{-10}	4.693×10^{-15}	1.051×10^{-15}
500	1.223×10^{-8}	5.462×10^{-11}	5.586×10^{-15}	5.276×10^{-15}
10,000	4.469×10^{-11}	1.054×10^{-13}	1.114×10^{-13}	1.115×10^{-13}

Table 2. The relative error of Clenshaw-Curtis quadrature rule for integral (30) [19].

k	N = 4	N = 7	N = 11	N = 16
20	3.710×10^{-3}	2.126×10^{-8}	2.846×10^{-13}	1.781×10^{-14}
100	3.016×10^{-3}	2.221×10^{-8}	1.473×10^{-13}	8.427×10^{-16}
500	2.924×10^{-3}	2.094×10^{-8}	1.408×10^{-13}	5.351×10^{-15}
10,000	3.047×10^{-3}	2.181×10^{-8}	1.836×10^{-13}	1.115×10^{-13}

Example 2. *Let integral*

$$\int_{-1}^{1} \frac{e^x \log(x-\alpha)e^{ikx}}{x-t} dx \tag{37}$$

Tables 3–5 represent results for relative error computed by Clenshaw-Curtis quadrature. As exact value we just have used that returned by the rule when a huge number of points is used.

Table 3. The relative error of Clenshaw-Curtis quadrature rule for integral (37) for $\alpha = -1$, $f(x) = e^x$, $t = 0.5$.

k	Exact Value	N = 4	N = 8	N = 10	N = 20
20	1.360346130460585 − 1.837213701909973i	3.505×10^{-6}	1.356×10^{-10}	1.744×10^{-13}	1.744×10^{-13}
100	0.528568077016834 + 2.007019282199925i	3.418×10^{-7}	8.530×10^{-12}	1.983×10^{-14}	0.00
500	2.032501926854849 + 0.510184343854610i	1.619×10^{-8}	3.974×10^{-13}	7.640×10^{-16}	5.297×10^{-17}
10,000	2.074653919328735 + 0.324969073545833i	6.131×10^{-11}	1.418×10^{-15}	2.114×10^{-16}	4.237×10^{-16}

Table 4. The relative error of Clenshaw-Curtis quadrature rule for integral (37) for $\alpha = 1$, $f(x) = e^x$, $t = 0.5$.

k	Exact Value	N = 4	N = 8	N = 10	N = 20
20	11.034821521905808 + 12.628898944623328i	1.144×10^{-6}	2.364×10^{-11}	1.232×10^{-13}	2.118×10^{-16}
100	−16.418938229588949 + 1.005287081095468i	3.020×10^{-8}	1.064×10^{-12}	3.356×10^{-15}	2.620×10^{-16}
500	−7.387722497395380 + 14.855177327546180i	3.485×10^{-9}	8.256×10^{-14}	2.394×10^{-16}	1.197×10^{-16}
10,000	−6.063084167285699 + 15.515830521473685i	1.132×10^{-11}	2.871×10^{-16}	1.192×10^{-16}	1.192×10^{-16}

Table 5. The relative error of Clenshaw-Curtis quadrature rule for integral (37) for $\alpha = 0$, $f(x) = e^x$, $t = 0.5$.

k	Exact Value	N = 4	N = 8	N = 10	N = 20
20	−1.928049736402945 + 2.990487262985703i	3.066×10^{-6}	8.567×10^{-11}	2.780×10^{-13}	1.248×10^{-16}
100	−0.934970743093483 − 3.460743549362822i	1.163×10^{-7}	2.942×10^{-12}	6.977×10^{-15}	2.770×10^{-16}
500	−3.485804022702049 − 0.864498281620865i	4.687×10^{-9}	1.177×10^{-13}	2.764×10^{-16}	9.274×10^{-17}
10,000	−3.547102638652960 − 0.555272021948841i	1.174×10^{-11}	2.473×10^{-16}	1.274×10^{-16}	3.092×10^{-17}

Example 3. Let the integral be

$$\int_{-1}^{1} \frac{\cos(x)\log(x-\alpha)e^{ikx}}{x-t} dx \tag{38}$$

Tables 6–8 represent results for relative error computed by Clenshaw-Curtis quadrature. As exact value is calculated by using the rule for large number of points.

Table 6. The relative error of Clenshaw-Curtis quadrature rule for integral (38) for $\alpha = -1$, $f(x) = \cos(x)$, $t = 0.8$.

k	Exact Value	N = 4	N = 8	N = 10	N = 20
20	0.498125821203593 − 1.281802863555419i	6.031×10^{-6}	2.150×10^{-10}	3.026×10^{-13}	9.449×10^{-16}
100	1.264215353181015 − 0.141780191524840i	5.295×10^{-7}	1.348×10^{-11}	3.197×10^{-14}	3.728×10^{-16}
500	1.090289998226562 − 0.675652244977728i	2.584×10^{-8}	6.432×10^{-13}	1.596×10^{-15}	1.935×10^{-16}
10,000	−1.283945795748914 + 0.084367340279936i	9.738×10^{-11}	2.749×10^{-15}	3.649×10^{-16}	1.186×10^{-16}

Table 7. The relative error of Clenshaw-Curtis quadrature rule for integral (38) for $\alpha = 1$, $f(x) = \cos(x)$, $t = 0.8$.

k	Exact Value	N = 4	N = 8	N = 10	N = 20
20	5.342145332192533 + 5.729353825896764i	2.057×10^{-6}	4.505×10^{-11}	2.427×10^{-13}	9.620×10^{-16}
100	−2.621138174403697 + 7.318981197518284i	4.131×10^{-8}	1.918×10^{-12}	6.183×10^{-15}	3.657×10^{-16}
500	0.622301278817091 + 7.666316541113909i	6.007×10^{-9}	1.558×10^{-13}	4.418×10^{-16}	1.291×10^{-16}
10,000	3.064017684660896 − 7.095233976390074i	1.886×10^{-11}	4.632×10^{-16}	1.284×10^{-16}	1.284×10^{-16}

Table 8. The relative error of Clenshaw-Curtis quadrature rule for integral (38) for $\alpha = 0$, $f(x) = \cos(x)$, $t = 0.8$.

k	Exact Value	N = 4	N = 8	N = 10	N = 20
20	−0.112551138814753 + 0.430514461423602i	2.401×10^{-5}	6.824×10^{-10}	2.176×10^{-12}	3.111×10^{-15}
100	−0.477210698149339 + 0.058677959322354i	8.618×10^{-7}	2.188×10^{-11}	5.266×10^{-14}	4.883×10^{-16}
500	−0.417276484590423 + 0.257429625049396i	3.407×10^{-8}	8.609×10^{-13}	7.249×10^{-16}	1.382×10^{-15}
10,000	0.487266314746835 − 0.032032920039315i	8.567×10^{-11}	2.205×10^{-15}	6.556×10^{-16}	4.626×10^{-16}

5. Conclusions

Clearly, Tables 1–8 illustrate the relative error of the Clenshaw-Curtis quadrature taken as $\frac{|I_{N+1}^\alpha[f] - I^\alpha[f]|}{|I^\alpha[f]|}$. We can see that for proposed Clenshaw-Curtis quadrature based on Hermite interpolation polynomial, with small value of points higher precision of the numerical results of integrals is obtained in $O(N \log N)$ operations. Furthermore these tables show that more accurate results can be obtained as k increases with fixed value of N. Conversely, more accurate approximation can be achieved as N increases but k is fixed. Moreover, Tables demonstrate that results successfully satisfy the analysis derived in Section 3.

Author Contributions: The Authors have equally contributed to this paper.

Funding: This work was supported by National Science Foundation of China (No. 11771454) and the Mathematics and Interdisciplinary Sciences Project of Central South University.

Conflicts of Interest: The authors declare no conflict of interest.The funders had no role in the design of the study; in the collection, analyses, or interpretation of data; in the writing of the manuscript, or in the decision to publish the results.

References

1. Karczmarek, P.; Pylak, D.; Sheshko, M.A. Application of Jacobi polynomials to approximate solution of a singular integral equation with Cauchy kernel. *Appl. Math. Comput.* **2006**, *181*, 694–707. [CrossRef]
2. Cuminato, N.A. On the uniform convergence of a perturbed collocation method for a class of Cauchy integral equations. *Appl. Numer. Math.* **1995**, *16*, 439–455. [CrossRef]
3. Eshkuvatov, Z.K.; Long, N.N.; Abdulkawi, M. Approximate solution of singular integral equations of the first kind with Cauchy kernel. *Appl. Math. Lett.* **2009**, *22*, 651–657. [CrossRef]
4. Ursell, F. Integral equations with a rapidly oscillating kernel. *J. Lond. Math. Soc.* **1969**, *1*, 449–459. [CrossRef]
5. Polyanin, A.D.; Manzhirov, A.V. *Handbook of Integral Equations*; CRC Press: Boca Raton, FL, USA, 1998. [CrossRef]
6. Olver, S. Numerical Approximation of Highly Oscillatory Integrals. Ph.D. Thesis, University of Cambridge, Cambridge, UK, 2008.
7. Iserles, A.; Norsett, S.P. On quadrature methods for highly oscillatory integrals and their implementation. *BIT Numer. Math.* **2004**, *44*, 755–772. [CrossRef]
8. Trefethen, L.N. Is Gauss quadrature better than Clenshaw Curtis? *SIAM Rev.* **2008**, *50*, 67–87. [CrossRef]
9. Trefethen, L.N. Chebyshev Polynomials and Series, Approximation theory and approximation practice. *Soc. Ind. Appl. Math.* **2013**, *128*, 17–19.
10. Dominguez, V.; Graham, I.G.; Smyshlyaev, V.P. Stability and error estimates for Filon-Clenshaw-Curtis rules for highly oscillatory integrals. *SIMA J. Numer. Anal.* **2011**, *31*, 1253–1280. [CrossRef]
11. Dominguez, V.; Graham, I.G.; Kim, T. Filon Clenshaw Curtis rules for highly oscillatory integrals with algebraic singularities and stationary points. *SIMA J. Numer. Anal.* **2013**, *51*, 1542–1566. [CrossRef]
12. Wang, H.; Xiang, S. Uniform approximations to Cauchy principal value integrals of oscillatory functions. *Appl. Math. Comput.* **2009**, *215*, 1886–1894. [CrossRef]
13. He, G.; Xiang, S. An improved algorithm for the evaluation of Cauchy principal value integrals of oscillatory functions and its application. *J. Comput. Appl. Math.* **2015**, *280*, 1–13. [CrossRef]
14. Xiang, S.; Chen, X.; Wang, H. Error bounds for approximation in Chebyshev points. *Numer. Math.* **2010**, *116*, 463–491. [CrossRef]
15. Setia, A. Numerical solution of various cases of Cauchy type singular integral equation. *Appl. Math. Comput.* **2014**, *230*, 200–207. [CrossRef]
16. Dominguez, V. Filon Clenshaw Curtis rules for a class of highly oscillatory integrals with logarithmic singularities. *J. Comput. Appl. Math.* **2014**, *261*, 299–319. [CrossRef]
17. Piessens, R.; Branders, M. On the computation of Fourier transforms of singular functions. *J. Comput. Appl. Math.* **1992**, *43*, 159–169. [CrossRef]

18. Chen, R. Fast computation of a class of highly oscillatory integrals. *Appl. Math. Comput.* **2014**, *227*, 494–501. [CrossRef]
19. Fang, C. Efficient methods for highly oscillatory integrals with weak and Cauchy singularities. *Int. J. Comput. Math.* **2016**, *93*, 1597–1610. [CrossRef]
20. Tsalamengas, J.L. Gauss Jacobi quadratures for weakly, strongly, hyper and nearly singular integrals in boundary integral equation methods for domains with sharp edges and corners. *J. Comput. Phys.* **2016**, *325*, 338–357. [CrossRef]
21. Liu, G.; Xiang, S. Clenshaw Curtis-type quadrature rule for hypersingular integrals with highly oscillatory kernels. *Appl. Math. Comput.* **2019**, *340*, 251–267. [CrossRef]
22. Boyd, J.P. *Chebyshev and Fourier Spectral Methods*; Courier Corporation: The North Chelmsford, MA, USA, 2001.
23. Trefethen, L.N. Chebfun. Available online: https://www.chebfun.org (accessed on 27 April 2019).
24. Oliver, J. The numerical solution of linear recurrence relations. *Numer. Math.* **1968**, *11*, 349–360. [CrossRef]
25. Xiang, S.; He, G.; Cho, Y.N. On error bounds of Filon Clenshaw Curtis quadrature for highly oscillatory integrals. *Adv. Comput. Math.* **2015**, *41*, 573–597. [CrossRef]

© 2019 by the authors. Licensee MDPI, Basel, Switzerland. This article is an open access article distributed under the terms and conditions of the Creative Commons Attribution (CC BY) license (http://creativecommons.org/licenses/by/4.0/).

Article

Model for the Evaluation of an Angular Slot's Coupling Impedance

Dario Assante [1,*] and Luigi Verolino [2]

[1] Facoltà di Ingegneria, Università Telematica Internazionale Uninettuno, 00186 Roma, Italy
[2] Dipartimento di Ingegneria Elettrica e delle Tecnologie dell'Informazione, Università degli Studi di Napoli Federico II, 80138 Napoli, Italy; verolino@unina.it
* Correspondence: d.assante@uninettunouniversity.net

Received: 25 April 2019; Accepted: 11 May 2019; Published: 22 May 2019

Abstract: In high energy particle accelerators, a careful modeling of the electromagnetic interaction between the particle beam and the structure is essential to ensure the performance of the experiments. Particular interest arises in the presence of angular discontinuities of the structure, due to the asymmetrical behavior. In this case, semi-analytical models allow one to reduce the computational effort and to better understand the physics of the phenomena, with respect to purely numerical models. In the paper, a model for analyzing the electromagnetic interaction between a traveling charge particle and a perfectly conducting angular slot of a negligible thickness is discussed. The particle travels at a constant velocity along a straight line parallel to the axis of symmetry of the strip. The longitudinal and transverse coupling impedances are therefore evaluated for a wide range of parameters.

Keywords: particle accelerator; coupling impedance; dual integral equations

1. Introduction

Recent discoveries in high-energy particle accelerators are connected to the possibility to reach higher level of energies in the experiments [1]. One of the main limitations to the involved energy, that is to say to the current of the beam, is the instability of the particle due to the electromagnetic interaction with the surrounding structures [2]. The synthetic design parameter commonly adopted in literature to describe the electromagnetic interaction between a traveling particle and a structure is the coupling impedance [3–5]. This parameter is proportional to the energy lost by the traveling charge due to the interaction with the scattered fields produced by the surrounding structures. Equivalently, it is proportional to the energy that has to be spent to keep its speed constant, neglecting the slowing effect of the surrounding structures. For structures invariant along the charge traveling direction, a per-unit-length coupling impedance has to be introduced [4], whose longitudinal and transverse components can be defined as

$$Z_{\parallel}(r,\varphi,k) = -\frac{1}{q}\frac{1}{L}\int_{-L/2}^{L/2} E_z(r,\varphi,z,\omega) e^{jkz/\beta} dz \qquad Z_{\perp}(r,\varphi,k) = \frac{1}{k}\nabla_{\perp} Z_{\parallel}(r,\varphi,k), \qquad (1)$$

where L is an unitary length, $E_z(r,\varphi,z,\omega)$ the x-component of the electric field in the frequency domain, k the wavenumber, and the charge q is moving at constant velocity $v = \beta c$ along the z axis. The second equation in (1) is known as the Panofski–Wenzel theorem [6].

The research of new shapes of cavities with proper coupling impedances is actually of high interest for the design of even more efficient particle accelerators [7–9]. Nowadays, powerful tools allow performing the electromagnetic numerical analysis of complex structures [10]. However, analytical or semi-analytical solutions still play a valuable role in this field, enabling to better understand the physics of some phenomena. Modal analysis is often adopted for close structures [11,12], diffractive methods for high-frequency solutions, and integral formulations for open geometries or in the presence of edges [13,14].

Most of the studies related to the coupling impedance consider, in a cylindrical reference system, axially symmetric geometries. This choice is both because they represent most of the structures of interest and because the symmetry allows finding the solution with less effort or even in a semi-analytical form. In this paper, we want to analyze the interaction of a particle with a axially asymmetric structure, in particular an angular slot, as shown in Figure 1. This configuration is representative of particle accelerator components that break the axial symmetry. The proposed method is quite general and can be adopted for a wide class of scattering and diffraction problems [15–20]. It can be easily generalized and adopted to analyze similar geometries.

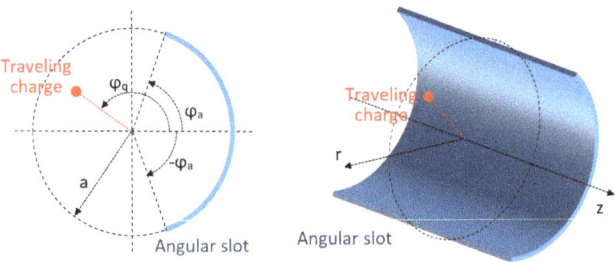

Figure 1. Geometry of the problem.

The problem is formulated in the particle frame at first. Assuming the geometry invariant along the traveling direction, in the particle frame a stationary model is adoptable, this simplifies the formulation and the solution. In order to evaluate the coupling impedance, the electromagnetic quantities are then obtained in the slot frame by means of Lorentz transforms.

The angular slot is assumed to be perfectly conductive; this is a common choice in the literature. Although the finite conductivity of the strip can be taken into account with some complications, neglecting it does not have a concrete effect on the validity of the analysis. The validity of such a choice is discussed more in detail in the last section of the paper.

The primed notation is adopted to identify the quantities in the particle frame, the unprimed notation in the slot frame.

The paper is composed of 6 sections: after this introduction, in the next section the problem is formulated in the particle frame and a methodology for computing the unknown current density is presented. In the following sections, the electromagnetic fields are evaluated in the particle and in the structure frame and then the coupling impedance is estimated. Then, some numerical results are presented. Finally, the conclusions are discussed.

2. Formulation of the Problem in the Particle Frame

In this section let us consider the geometry shown in Figure 1: a perfectly conducting angular slot $\mathbb{S} = \{r = a, |\varphi| \leq \varphi_a, z\}$ at distance a from the axis and covering an angular sector of $2\varphi_a$. A travelling charge q moves parallel to the slots's axis, placed at (r_q, φ_q), at constant speed $v = \beta c$, c being the speed of light in free space.

The problem is formulated in term of integral equations and its solution is reduced to the resolution of a linear system.

The electromagnetic interaction between the particle and the structure can be easily formulated and solved in the particle frame, being an electrostatic model adequate for such a problem. Once the electromagnetic quantities are computed, their values in the slot frame can be obtained by means of Lorentz transforms.

The electrostatic potential produced by the charge is

$$V'_q = \frac{q}{4\pi\varepsilon_0 \sqrt{r'^2 + r_q^2 - 2r'r_q \cos(\varphi' - \varphi_q) + z'^2}},\qquad(2)$$

while the potential produced by the induced charge density $\sigma'(\varphi', z')$ on the slot can be expressed as

$$V' = \frac{a}{4\pi\varepsilon_0} \int_S \frac{\sigma'(\varphi_0, z_0)\, d\varphi_0 dz_0}{\sqrt{r'^2 + a^2 - 2r'a \cos(\varphi' - \varphi_0) + (z' - z_0)^2}}.\qquad(3)$$

Being a perfectly conducting slot, the boundary condition to be verified is that the tangential components of the electric field vanishes on the slot. This corresponds to impose that

$$V'(r' = a, \varphi', z) + V'_q(r' = a, \varphi', z) = 0 \qquad(4)$$

for every $(\varphi', z') \in \mathbb{S}$.

Considering Equations (2) and (3), the boundary condition leads to

$$\int_S \frac{\sigma'(\varphi_0, z_0)\, d\varphi_0 dx_0}{\sqrt{2a^2 - 2a^2 \cos(\varphi' - \varphi_0) + (z' - z_0)^2}} = -\frac{q/a}{\sqrt{a^2 + r_q^2 - 2ar_q \cos(\varphi' - \varphi_q) + z'^2}} \qquad(5)$$

From this equation, it is worth noting to observe that there is complete induction on the slot for such a kind of geometry. In fact, by multiplying Equation (5) for the denominator of its second member and performing a limit for z going toward $+\infty$, it is possible to obtain that

$$\int_S \sigma'(\varphi_0, z_0)\, a\, d\varphi_0 dz_0 = -q. \qquad(6)$$

This result will be usefully employed later on in the computation of the coupling impedance.

In order to solve the problem it is necessary to recall a relevant integral ([21] R6.616.4)

$$\frac{\pi}{\sqrt{D^2 + Z^2}} = \int_{-\infty}^{+\infty} K_0(Dw)\, e^{-jwZ} dw, \qquad(7)$$

and its derived form

$$\frac{\pi Z}{(D^2 + Z^2)^{3/2}} = j \int_{-\infty}^{+\infty} w K_0(Dw)\, e^{-jwZ} dw. \qquad(8)$$

Then it is useful to introduce a spatial Fourier transform along the z axis, namely

$$\tilde{\sigma}'(\varphi, w) = \frac{1}{2\pi} \int_{-\infty}^{+\infty} \sigma'(\varphi, z) e^{jwz} dz. \qquad(9)$$

By using Equation (7) on the integral Equation (5) and applying the inverse Fourier transform to both members, with some manipulations it becomes

$$\int_{-\varphi_a}^{\varphi_a} \tilde{\sigma}'(\varphi_0, w) K_0 \left(\sqrt{2a}\, w \sqrt{1 - \cos(\varphi' - \varphi_0)} \right) d\varphi_0 =$$

$$= -\frac{q}{2\pi a} K_0 \left(a\, w \sqrt{1 + (r_q/a)^2 - 2(r_q/a) \cos(\varphi' - \varphi_q)} \right). \quad (10)$$

Equation (10) has to be verified on $\mathbb{S}$, whereas the induced charge density vanishes outside the angular slot, namely

$$\int_{-\infty}^{+\infty} \tilde{\sigma}'(\varphi_0, w)\, e^{-jwz} dw = 0. \quad (11)$$

Equations (10) and (11) constitute a dual system of integral equation with respect to the induced current density. An efficient solution of such a problem can be obtained by representing the unknown in terms of Neuman series [22]:

$$\sigma'(\varphi', z') = \begin{cases} -\dfrac{q}{2\pi a\, \varphi_a} \sum\limits_{n=0}^{\infty} \sigma_n(z') \dfrac{T_n(\varphi'/\varphi_a)}{\sqrt{1 - (\varphi'/\varphi_a)^2}}, & |\varphi| \le \varphi_a \\ 0, & |\varphi| > \varphi_a \end{cases} \quad (12)$$

where $T_n(\cdot)$ is the Chebychev polynomial of order n [23,24]. Such polynomials exhibit several relevant properties [25,26] that can be adopted for the solution of some classes of electromagnetic problems. According to Equation (9), then the induced current density in the transformed domain is

$$\tilde{\sigma}'(\varphi', w) = -\frac{q}{2\pi a\, \varphi_a} \sum_{n=0}^{\infty} \tilde{\sigma}_n(w) \frac{T_n(\varphi'/\varphi_a)}{\sqrt{1 - (\varphi'/\varphi_a)^2}}. \quad (13)$$

With such a normalization, dimensionless expansion coefficients are obtained in the transformed domain.

The chosen representation is a form of the more generic Neuman series, particularized for this problem [27]. The chosen representation automatically matches the right edge behavior, this regularizes the method and reduces the number of required coefficients.

In addition, the chosen current density representation already satisfies Equation (11). So, by substituting it into the remaining Equation (10), it is found that

$$\sum_{n=0}^{\infty} \tilde{\sigma}_n(w) \int_{-\varphi_a}^{\varphi_a} \frac{T_n(\varphi'/\varphi_a)}{\sqrt{1 - (\varphi'/\varphi_a)^2}} K_0 \left(\sqrt{2a}\, w \sqrt{1 - \cos(\varphi' - \varphi_0)} \right) d\varphi_0 =$$

$$= \varphi_a K_0 \left(a\, w \sqrt{1 + (r_q/a)^2 - 2(r_q/a) \cos(\varphi' - \varphi_q)} \right). \quad (14)$$

This equation has to be verified for every $|\varphi'| \le \varphi_a$. In order to impose this condition, Equation (14) is projected on the same basis functions adopted for the representation of the current density (Galerkin scheme). This converts Equation (14) in the linear system

$$\mathbf{A}\tilde{\boldsymbol{\sigma}} = \mathbf{b}, \quad (15)$$

where $\tilde{\boldsymbol{\sigma}}$ is the vector of the unknowns $\tilde{\sigma}_n$, $\mathbf{A}$ is a symmetric matrix whose coefficients, obtained with some trivial manipulations and changes of variable as

$$A_{nm} = \int_0^{\pi} \int_0^{\pi} K_0 \left(2a\, w \left| \sin\left(\frac{\varphi_a}{2} (\cos\psi_0 - \cos\psi') \right) \right| \right) \cos(m\psi_0) \cos(n\psi')\, d\psi_0 d\psi' \quad (16)$$

and b the known term vector defined as

$$b_m = \int_0^\pi K_0 \left(a\, w \sqrt{1 + (r_q/a)^2 - 2(r_q/a) \cos\left(\varphi_a \cos\psi' - \varphi_q\right)} \right) \cos\left(m\psi'\right) d\psi'. \tag{17}$$

It is worth noting that for $r_q = 0$, that is to say when the particle is in the axis of the slot, all the coefficients b_m are zero but the first one, that is $b_0 = \pi K_0(a\, w)$.

3. Electromagnetic Fields in the Slot Frame

In order to complete the problem formulation, it is proper to express the electromagnetic fields in the slot frame, too. This can be realized by applying the Lorentz transforms to the fields computed in the previous section in particle frame.

Let us consider at first the z component of the electric field. The contribution provided by the traveling charge in the frequency domain is well known and is

$$E_{z,q} = \frac{jq\kappa\zeta_0}{2\pi\beta\gamma} e^{-jzk/\beta} K_0\left(\kappa\sqrt{r^2 + r_q^2 - 2rr_q \cos(\varphi)}\right), \tag{18}$$

where $\gamma = 1/\sqrt{1-\beta^2}$ is the Lorentz factor, $\kappa = k/(\beta\gamma)$, and $\zeta_0 = \sqrt{\mu_0/\varepsilon_0}$ is the characteristic impedance of free space.

The contribution produced by the induced current density on the slot can be obtained with some manipulations as function of the representation coefficients σ_n.

In the particle frame, starting from Equation (3) it is possible to obtain

$$e_z'(r', \varphi', z') = \frac{a}{4\pi\varepsilon_0} \int_S \frac{\sigma'(\varphi_0, z_0)\,(z'-z_0)\, d\varphi_0 dz_0}{\left[r'^2 + a^2 - 2r'a\cos(\varphi'-\varphi_0) + (z'-z_0)^2\right]^{3/2}}. \tag{19}$$

Lorentz transforms are now applied to obtain the electric field in the slot frame. In this specific case they are

$$e_z' = e_z,\ \sigma' = \sigma\gamma,\ r' = r,\ \varphi' = \varphi,\ z' = \gamma(z - vt). \tag{20}$$

Applying these transforms to Equation (19), it is found that

$$e_z(r, \varphi, z, t) = \frac{a\gamma}{4\pi\varepsilon_0} \int_S \frac{\sigma(\varphi_0, z_0)\,(\gamma(z-vt)-z_0)\, d\varphi_0 dz_0}{\left[r^2 + a^2 - 2ra\cos(\varphi-\varphi_0) + (\gamma(z-vt)-z_0)^2\right]^{3/2}}. \tag{21}$$

By means of Equation (8) and applying a spatial Fourier transform according to Equation (9), it is found that

$$e_z(r, \varphi, z, t) = \frac{ja\,\gamma}{2\pi\varepsilon_0} \int_{-\varphi_a}^{+\varphi_a} \int_{-\infty}^{+\infty} \tilde{\sigma}(\varphi_0, w) w e^{jw\gamma vt} K_0\left(w\sqrt{r^2 + a^2 - 2ra\cos(\varphi-\varphi_0)}\right) e^{-jw\gamma z} d\varphi_0 dw. \tag{22}$$

Finally, by performing a time Fourier transform and then the integral on w, the required field is finally found as

$$E_z(r, \varphi, z, \omega) = \frac{ja\, k\zeta_0}{\beta^2} e^{-jzk/\beta} \int_{-\varphi_a}^{+\varphi_a} \tilde{\sigma}(\varphi_0, \kappa) K_0\left(\kappa\sqrt{r^2 + a^2 - 2ra\cos(\varphi-\varphi_0)}\right) d\varphi_0. \tag{23}$$

The last integral can be performed by substituting the representation of the current density (13), leading to the very simple expression of the z-component of the electric field in the slot frame:

$$E_z(r, \varphi, z, \omega) = -\frac{jqk\zeta_0}{2\pi\varphi_a\beta^2} e^{-jzk/\beta} \sum_{n=0}^{\infty} \tilde{\sigma}_n(\kappa) b_n(r, \varphi, \kappa), \tag{24}$$

where coefficients $b_n(r, \varphi, \kappa)$ have the same expression of Equation (29) but are computed as the generic point (r, φ) and for $w = \kappa$.

With a similar procedure, all the other fields can be expressed. For instance, it is possible to easily find that the charge density induced on the slot is

$$\sigma(\varphi, z, \omega) = -\frac{q}{a\varphi_a\gamma} \sum_{n=0}^{\infty} \tilde{\sigma}_n(\kappa) \frac{T_n(\varphi/\varphi_a)}{\sqrt{1 - (\varphi/\varphi_a)^2}}. \tag{25}$$

4. Coupling Impedance

In order to compute the coupling impedance, just the longitudinal component of the electric field produced by the induced currents is required, since the electric field produced by the traveling charge does not contribute to the coupling impedance.

So, given Equation (24), the per-unit-length longitudinal coupling impedance (1) can be easily computed in a generic point in the transverse plane as

$$Z_{\parallel}(r, \varphi, k) = \frac{jk\zeta_0}{2\pi\varphi_a\beta^2} \sum_{n=0}^{\infty} \tilde{\sigma}_n(\kappa) b_n(\kappa, r, \varphi). \tag{26}$$

It is worth noting that, since the matrix and the known term vector in Equation (15) are purely real, all the unknown terms σ_n are real, too. So, the longitudinal coupling impedance is purely imaginary. This result is expected since there are not diffraction losses.

From Equation (26), by means of the definition (1) it is possible to find the expression of the transverse coupling impedance, that is

$$Z_{\perp}(r, \varphi, k) = \frac{jk\zeta_0}{2\pi\varphi_a\beta^2} \sum_{n=0}^{\infty} \tilde{\sigma}_n(\kappa) \left\{ \frac{\partial b_n(\kappa, r, \varphi)}{\partial r} \hat{r} + \frac{1}{r} \frac{\partial b_n(\kappa, r, \varphi)}{\partial \varphi} \hat{\varphi} \right\}. \tag{27}$$

In order to compute the transverse coupling impedance in a practical way, it is worth recalling the addition theorem for the Hankel functions

$$K_0(wR) = \sum_{p=-\infty}^{+\infty} (-1)^p I_p(\rho'w) K_p(\rho w) e^{jp(\varphi-\varphi')}, \tag{28}$$

being $R = \sqrt{\rho'^2 + \rho^2 - 2\rho'\rho\cos(\phi'-\phi)}$ and $\rho' \leq \rho$.

With some manipulations, it can be used to analytically compute the integral in Equation (17), expressing the coefficients b_n as series of products of Bessel functions, namely

$$b_m(r_q, \varphi_q, w) = \pi \begin{cases} I_0(r_q w) K_0(aw) + 2 \sum_{p=1}^{+\infty} (-1)^p I_p(r_q w) K_p(aw) J_0(p\varphi_a) \cos(p\varphi_q), & m = 0, \\ 2j^m \sum_{p=1}^{+\infty} (-1)^p I_p(r_q w) K_p(aw) J_m(p\varphi_a) \cos(p\varphi_q), & m \text{ even}, \\ 2j^m \sum_{p=1}^{+\infty} (-1)^p I_p(r_q w) K_p(aw) J_m(p\varphi_a) \sin(p\varphi_q), & m \text{ odd}. \end{cases} \tag{29}$$

Then, the derivatives required in Equation (27) can be expressed as

$$\frac{\partial b_n}{\partial r}(r,\varphi,\kappa) = 2\pi j^n \begin{cases} \frac{1}{2} I_1(r\kappa) K_0(a\kappa) + \\ + \sum_{p=1}^{+\infty} (-1)^p \left[\frac{p}{r} I_p(r\kappa) + I_{p+1}(r\kappa)\right] K_p(a\kappa) J_0(p\varphi_a) \cos(p\varphi), & m = 0, \\ \sum_{p=1}^{+\infty} (-1)^p \left[\frac{p}{r} I_p(r\kappa) + I_{p+1}(r\kappa)\right] K_p(a\kappa) J_n(p\varphi_a) \cos(p\varphi), & n \text{ even,} \\ \sum_{p=1}^{+\infty} (-1)^p \left[\frac{p}{r} I_p(r\kappa) + I_{p+1}(r\kappa)\right] K_p(a\kappa) J_n(p\varphi_a) \sin(p\varphi), & n \text{ odd.} \end{cases} \quad (30)$$

and

$$\frac{1}{r}\frac{\partial b_n}{\partial \varphi}(r,\varphi,\kappa) = \frac{2\pi j^n}{r} \begin{cases} \sum_{p=1}^{+\infty} (-1)^{(p+1)} p I_p(r\kappa) K_p(a\kappa) J_n(p\varphi_a) \sin(p\varphi), & n \text{ even,} \\ \sum_{p=1}^{+\infty} (-1)^{(p+1)} p I_p(r\kappa) K_p(a\kappa) J_n(p\varphi_a) \cos(p\varphi), & n \text{ odd.} \end{cases} \quad (31)$$

Among all the possible positions in the transverse plane, a relevant one is when the particle is in the center of the slot, namely $r = 0$. In fact, in a particle accelerator, the particle is supposed to travel along the radial axis of the structure, or eventually in a position very close to it. In this case, as previously stated, all the coefficients $b_n(\kappa, 0)$ vanishes but the one for $n = 0$.

So the longitudinal coupling impedance Equation (26) assumes the very simple expression

$$Z_\parallel(r,\varphi,k) = \frac{jk\zeta_0}{2\pi \varphi_a \beta^2} \tilde{\sigma}_0(\kappa) b_0(\kappa, 0). \quad (32)$$

Then, performing the limits for r going to zero of Equations (30) and (31), the coefficients of Equation (27) reduces to

$$\frac{\partial b_n}{\partial r}(r,\varphi,\kappa) = -\pi j^n K_1(a\kappa) J_n(\varphi_a) \begin{cases} \cos(\varphi), & n \text{ even,} \\ \sin(\varphi), & n \text{ odd.} \end{cases} \quad (33)$$

and

$$\frac{1}{r}\frac{\partial b_n}{\partial \varphi}(r,\varphi,\kappa) = \pi j^n K_1(a\kappa) J_n(\varphi_a) \begin{cases} \sin(\varphi), & n \text{ even,} \\ \cos(\varphi), & n \text{ odd.} \end{cases} \quad (34)$$

So the transverse coupling impedance (27) can be easily expressed in Cartesian coordinates and becomes

$$Z_\perp(r,\varphi,k) = -\frac{jk\zeta_0}{2\varphi_a \beta^2} K_1(a\kappa) \sum_{n=0}^{\infty} j^n \tilde{\sigma}_n(\kappa) J_n(\varphi_a) \hat{x}. \quad (35)$$

Such a result is coherent with the physics of the problem since, due to symmetry reasons, for such particle position it is expected that the transverse coupling impedance is along the bisector of the angular slot.

5. Numerical Results

Some numerical results are presented in this section, in order to discuss the efficiency of the proposed method. In all the simulations, the shape of the angular slot is $a = 1$ cm, $\varphi_a = 60°$. A Simpson rule with an adaptive spacing is adopted to compute the matrix coefficients (16), while a Gaussian quadrature algorithm is used for the coefficients (17). Since the kernel of the terms in Equation (16) exhibits a logarithmic singularity and gives rise to computational problems, proper numerical manipulations have to be introduced to navigate the problem. The adopted solution is discussed in the Appendix A. At first, the behavior of the coefficients σ_n is shown for different values of the frequency and of the distance between the particle and the structure. In Figure 2a the absolute values of expansion coefficients are shown for different frequencies. The particle is in the center of the axis, as that is the most realistic case in practice. At lower frequencies, the coefficients' amplitudes quickly decrease, with few of them being enough to properly represent the current density: for $a\kappa = 0.01$, the third coefficient is already four orders of magnitude lower than the first one. At higher frequencies, the amplitude of the coefficients decreases more slowly, so an higher number of coefficients is required, as expected.

In Figure 2b the coefficients are shown in case of offset of the particle beam. As expected, the amplitude of the coefficients increases as the distance of the particle from the structure decreases. Additionally, while the odd coefficients vanish in case of centred particle, they grow proportionally to the particle offset.

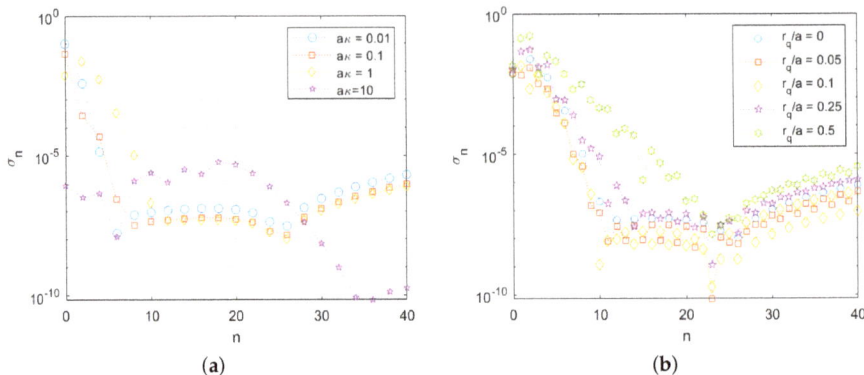

Figure 2. Absolute values of the expansion coefficients at $\beta\gamma = 1$ for: (**a**) different frequencies ($r_q = 0$), (**b**) different offsets of the particle ($a\kappa = 1$, $\varphi_q = 30°$).

Then, in Figure 3 we show the behavior of the current density induced on the angular slot as in Equation (25), for different values of the frequency, normalized with respect to the charge. The adopted coefficients are the same as in Figure 2a. As expected, at low frequencies induced currents have a very flat behavior, just exhibiting a divergence at the boundaries. As the frequency grows, the behavior of the current density is more variable even in the middle of the angular slot. The results have been successfully validated with a finite element tool, providing a very good correspondence in the middle of the structure and being unable to reproduce the proper divergent boundary behavior, as expected.

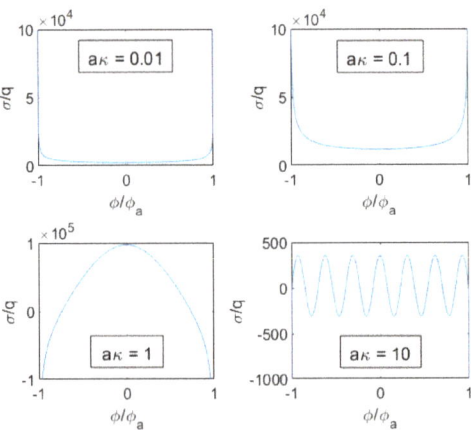

Figure 3. Behavior of the current density induced on the structure ($r_q = 0$, $\beta\gamma = 1$).

Finally, the coupling impedance is discussed. In Figure 4a, the longitudinal coupling impedance is shown, as a function of the frequency and of the particle speed. As already discussed, in practice it is a coupling reactance, the real part being null. The shape is slightly influenced by the particle speed. As expected, the coupling impedance drops to negligable values for high frequencies.

In Figure 4b, the transverse coupling impedance (reactance) is shown, as function of the frequency and of the particle speed, with similar considerations with respect to the longitudinal impedance.

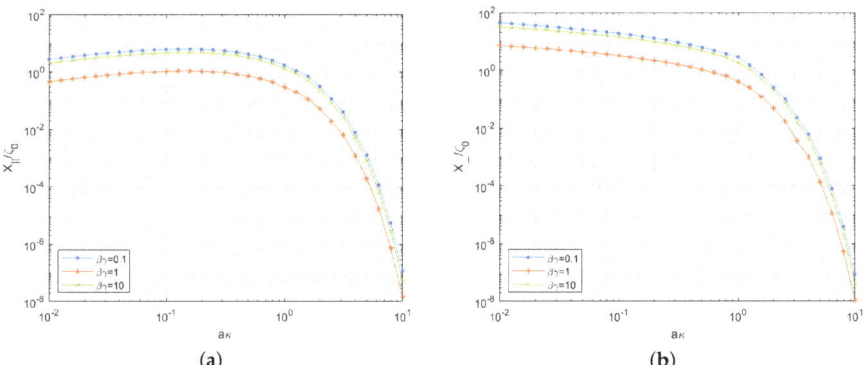

Figure 4. Normalized per unit length (**a**) longitudinal and (**b**) transverse coupling reactance for different frequencies ($r_q = 0$).

6. Discussion

A method for the evaluation of the coupling impedance of a particle travelling parallel close to a perfectly conducting angular slot has been presented. The method is accurate and effective, and can be easily generalized to similar geometries exhibiting angular variation in cylindrical coordinates.

As stated in the introduction, the angular slot is assumed to be perfectly conducting. This assumption is widely accepted in the literature on particle accelerator cavity design, for several reasons. First of all, the coupling impedance mainly takes into account the electromagnetic interaction between the particle and the surrounding structure, which is often mainly connected to structure shape. In this sense, the real part of the coupling impedance aims to take into account the diffractive losses. For this reason, most of the scientific paper consider the structure as perfectly conducting. In most of the cases where the conductivity of the structure is taken into account, this aspect is usually added to the perfectly conducting model with perturbative approaches and its effect usually smooths some rough behaviors but does not produce relevant changes. For this reason, the study of structures with perfectly conducting walls is often considered a valuable analysis, even if the inclusion of the finite conductivity is of course an added value. Regarding our specific problem, most of the analysis does not lose validity, even when adding the finite conductivity of the slot. Since the problem is formulated in the particle frame at first, the model is stationary regardless of whether the slot's conductivity is finite or not. So the first part of the paper is not affected. After applying the Lorentz transforms, once the current density is found in the strip frame, it is possible to evaluate the resistive power dissipation on the slot. Finally, in real particle accelerators, despite whether the particle speed is close to the light speed, the charge is extremely small and the current is not very high, usually hundreds of mA at most. Therefore, the current densities in practical cases are not huge.

The angular slot and the travelling charge are placed in an open space. In a particle accelerator, all the components are closed in a metallic pipe, which is necessary to maintain a vacuum. With respect to the proposed formulation, it is possible to add the presence of the pipe by partially changing the kernel of Equations (16) and (17). However, such a change just introduces some poles in the kernel, connected to the pipe resonances, independent of the asymmetry of the angular slot. Since the aim of the paper is to propose a method to deal with angularly asymmetric structures, just focussing on that and not mixing different phenomena, the external pipe has been neglected.

The proposed method has proven to be accurate and the series is quickly convergent. It is suitable to analyze structures with angular asymmetry. The obtained results can be used to benchmark numerical solutions or as reference geometry for more complex structures typical of particle accelerators.

Author Contributions: Conceptualization, D.A. and L.V.; methodology, D.A. and L.V.; software, D.A.; validation, D.A. and L.V.

Funding: This research received no external funding.

Conflicts of Interest: The authors declare no conflict of interest.

Appendix A. Computation of the Linear System

For low values of the argument, the kernel of the integral of Equation (16) exhibits a logarithmic singularity, namely

$$K_0(z) \cong -\log(z/2) - \gamma_0, \tag{A1}$$

with γ_0 being the Eulero–Mascheroni constant. For its efficient numerical computation, it is worth adopting the variational form

$$A_{nm} = A_{nm}^{\log} + A_{nm}^0, \tag{A2}$$

where

$$A_{nm}^0 = \int_0^\pi \int_0^\pi \left[K_0 \left(2a\, w \left| \sin\left(\frac{\varphi_a}{2} (\cos\psi_0 - \cos\psi') \right) \right| \right) + \right.$$
$$\left. + \log\left(a\, w \left| \sin\left(\frac{\varphi_a}{2} (\cos\psi_0 - \cos\psi') \right) \right| \right) + \gamma_0 \right] \cos(m\psi_0) \cos(n\psi')\, d\psi_0 d\psi'. \tag{A3}$$

The new integral in Equation (A3) has no singularities and can be numerically computed with minimal effort. Regarding the logarithmic, by means of the relevant expansion

$$\log|\sin(x/2)| = -\sum_{p=1}^\infty \frac{\cos px}{p} - \log 2, \tag{A4}$$

with some manipulations it can be easily found that

$$A_{nm}^{\log} = -\int_0^\pi \int_0^\pi \left[\log\left(a\, w \left| \sin\left(\frac{\varphi_a}{2} (\cos\psi_0 - \cos\psi') \right) \right| \right) + \gamma_0 \right] \cos(m\psi_0) \cos(n\psi')\, d\psi_0 d\psi' =$$

$$= \begin{cases} -\pi^2 [\log(a\, w/2) + \gamma_0], & m = n = 0 \\ \pi^2 \sum_{p=1}^\infty \dfrac{J_m(p/\varphi_a) J_n(p/\varphi_a)}{p}, & m+n \text{ odd}, \\ 0, & m+n \text{ even}. \end{cases} \tag{A5}$$

References

1. Lee, S.Y. *Accelerator Physics*; World Scientific Publishing: Singapore, 2018.
2. Kheifets, S.A.; Zotter, B.W. *Impedances and Wakes in High-Energy Particle Accelerators*; World Scientific: Singapore, 1998.
3. Heifets, S.A.; Kheifets, S.A. Coupling impedance in modern accelerators. *Rev. Mod. Phys.* **1993**, *63*, 631–673. [CrossRef]
4. Palumbo, L.; Vaccaro, V.G.; Zobov, M. Wake Fields and Impedance. 1995. Available online: https://arxiv.org/abs/physics/0309023 (accessed on 21 May 2019).
5. Chao, A.W.; Mess, K.H. *Handbook of Accelerator Physics and Engineering*; World Scientific: Singapore, 1998.
6. Panofsky, W.; Wenzel, W. Transverse deflection of charged particles in radiofrequency fields. *Rev. Sci. Instrum.* **1956**, *27*, 967–977. [CrossRef]
7. Niedermayer, U.; Eidam, L.; Boine-Frankenheim, O. Analytic modeling, simulation and interpretation of broadband beam coupling impedance bench measurements. *Nucl. Instrum. Methods Phys. Res. Sect. A Accel. Spectrom. Detect. Assoc. Equip.* **2015**, *776*, 129–143. [CrossRef]
8. Zannini, C. Electromagnetic Simulation of CERN Accelerator Components and Experimental Applications. 2013. Available online: http://inspirehep.net/record/1296518/ (accessed on 21 May 2019).
9. Campelo, J.E.; Ghini, J.; Salvant, B.; Argyropoulos, T.; Esteban Müller, J.; Ghini, J.; Lasheen, A.; Quartullo, D.; Salvant, B.; Shaposhnikova, E.; et al. An Extended SPS Longitudinal Impedance Model. 2015. Available online: http://inspirehep.net/record/1417277/ (accessed on 21 May 2019).
10. Fujita, K.; Kawaguchi, H. Time Domain Numerical Simulation Method based on EFIE and MFIE for Axis-Symmetric Structure Objects. In Proceedings of the Progress in Electromagnetic Research Symposium 2004, Pisa, Italy, 28–31 March 2004; pp. 627–630.
11. Kuehn, E. A mode-matching method for solving field problems in waveguide and resonator circuits. *Arch. Fuer Elektron. Und Uebertragungstechnik* **1973**, *27*, 511–518.
12. Legenkiy, M.N.; Butrym, A.Y. Method of mode matching in time domain. *Prog. Electromagn. Res.* **2010**, *22*, 257–283. [CrossRef]
13. Assante, D.; Davino, D.; Falco, S.; Schettino, F.; Verolino, L. Coupling impedance of a charge travelling in a drift tube. *IEEE Trans. Mag.* **2005**, *41*, 1924–1927. [CrossRef]
14. Assante, D.; Verolino, L. Efficient evaluation of the longitudinal coupling impedance of a plane strip. *Prog. Electromagn. Res.* **2012**, *26*, 251–265. [CrossRef]
15. Censor, D. Free-space relativistic low-frequency scattering by moving objects. *Prog. Electromagn. Res.* **2007**, *72*, 195–214. [CrossRef]
16. Censor, D. Application-oriented relativistic electrodynamics. *Prog. Electromagn. Res.* **2000**, *29*, 107–168. [CrossRef]
17. Idemen, M.; Alkumru, A. Influence of motion on the edge-diffraction. *Prog. Electromagn. Res.* **2008**, *6*, 153–168. [CrossRef]
18. Censor, D. Broadband spatiotemporal differential-operator representations for velocity depending scattering. *Prog. Electromagn. Res.* **2006**, *58*, 51–70. [CrossRef]
19. Sautbekov, S.S. Diffraction of plane wave by strip with arbitrary orientation of wave vector. *Prog. Electromagn. Res.* **2011**, *21*, 117–131. [CrossRef]
20. Handapangoda, C.C.; Premaratne, M.; Pathirana, P.N. Plane wave scattering by a spherical dielectric particle in motion: A relativistic extension of the mie theory. *Prog. Electromagn. Res.* **2011**, *112*, 349–379. [CrossRef]
21. Gradshteyn, I.S.; Ryzhik, I.M. *Table of Integrals, Series, and Products*, 7th ed.; Academic Press: New York, NY, USA, 2007.
22. Eswaran, K. On the solutions of a class of dual integral equations occurring in diffraction problems. *Proc. R. Soc. Lond.* **1990**, *A429*, 399–427. [CrossRef]
23. Cesarano, C. Generalized Chebyshev polynomials. *Hacet. J. Math. Stat.* **2014**, *43*, 731–740.
24. Cesarano, C. Identities and generating functions on Chebyshev polynomials. *Georgian Math. J.* **2012**, *19*, 427–440. [CrossRef]
25. Cesarano, C. Integral representations and new generating functions of Chebyshev polynomials. *Hacet. J. Math. Stat.* **2015**, *44*, 541–552. [CrossRef]

26. Cesarano, C.; Fornaro, F. A note on two-variable Chebyshev polynomials. *Georgian Math. J.* **2017**, *24*, 339–350. [CrossRef]
27. Assante, D.; Falco, S.; Lucido, M.; Panariello, G.; Schettino, F.; Verolino, L. Shielding effect of a strip of finite thickness. *Electr. Eng.* **2006**, *89*, 79–87. [CrossRef]

 © 2019 by the authors. Licensee MDPI, Basel, Switzerland. This article is an open access article distributed under the terms and conditions of the Creative Commons Attribution (CC BY) license (http://creativecommons.org/licenses/by/4.0/).

Article

Asymptotic Properties of Solutions of Fourth-Order Delay Differential Equations

Clemente Cesaro [1,*], Sandra Pinelas [2], Faisal Al-Showaikh [3] and Omar Bazighifan [4]

[1] Section of Mathematics, International Telematic University Uninettuno, CorsoVittorio Emanuele II, 39, 00186 Roma, Italy
[2] RUDN University, 6 Miklukho-Maklaya Str., 117198 Moscow, Russia; sandra.pinelas@gmail.com
[3] Department of Mathematics, Faculty of Science, University of Bahrain, Zallaq 1051, Bahrain; falshawaikh@uob.edu.bh
[4] Department of Mathematics, Faculty of Science, Hadhramout University, Hadhramout 50512, Yemen; o.bazighifan@gmail.com
* Correspondence: c.cesarano@uninettunouniversity.net

Received: 28 March 2019; Accepted: 25 April 2019; Published: 3 May 2019

Abstract: In the paper, we study the oscillation of fourth-order delay differential equations, the present authors used a Riccati transformation and the comparison technique for the fourth order delay differential equation, and that was compared with the oscillation of the certain second order differential equation. Our results extend and improve many well-known results for oscillation of solutions to a class of fourth-order delay differential equations. Some examples are also presented to test the strength and applicability of the results obtained.

Keywords: fourth-order; nonoscillatory solutions; oscillatory solutions; delay differential equations

1. Introduction

In this work, we consider a fourth-order delay differential equation

$$Lz + q(y) f(z(\sigma(y))) = 0, \tag{1}$$

where

$$Lz := \left[m_3(y) \left(m_2(y) \left[m_1(y) z'(y) \right]' \right)' \right]'.$$

We assume $m_i, q, \sigma \in C([y_0, \infty), \mathbb{R})$, $m_i(y) > 0$, $i = 1, 2, 3$, $\lim_{y \to \infty} \frac{m_3(y)}{m_1(y)} > 0$, $q > 0$, $\sigma(y) \leq y$ and $\lim_{y \to \infty} \sigma(y) = \infty$, $f \in C(\mathbb{R}, \mathbb{R})$, $f(u)/u \geq k > 0$ for $u \neq 0$.

By a solution of (1) we mean a function $z \in C((\sigma(y_z), \infty))$, which has the property $m_1(y) z'(y)$, $m_2(y) [m_1(y) z'(y)]'$, $m_3(y) \left(m_2(y) [m_1(y) z'(y)]' \right)' \in C^1[y_z, \infty)$, and satisfies (1) on $[y_z, \infty)$. We consider only those solutions z of (1) which satisfy $\sup\{|z(y)| : y \geq y_z\} > 0$, for all $y > y_z$. Such a solution is said to be oscillatory if it has arbitrarily large zeros and nonoscillatory otherwise.

The study of differential equations with deviating argument was initiated in 1918, appearing in the first quarter of the twentieth century as an area of mathematics that has since received a lot of attention. It has been created in order to unify the study of differential and functional differential equations. Since then, there has been much research activity concerning the oscillation of solutions of various classes of differential and functional differential equations. Many authors have contributed on various aspects of this theory, see ([1–9]) .

The problem of the oscillation of higher and fourth order differential equations have been widely studied by many authors, who have provided many techniques used for obtaining oscillatory criteria for higher and fourth order differential equations. We refer the reader to the related books (see [4,10–13]) and to the papers (see [11,14–18]). Because of the above motivating factors for the study of fourth-order differential equations, as well as because of the theoretical interest in generalizing and extending some known results from those given for lower-order equations, the study of oscillation of such equations has received a considerable amount of attention. For a systematic summary of the most significant efforts made as regards this theory, the reader is referred to the monographs of [19–22].

Especially, second and fourth order delay differential equations are of great interest in biology in explaining self-balancing of the human body and in robotics in constructing biped robots.

One of the traditional tools in the study of oscillation of equations which are special cases of (1) has been based on a reduction of order and the comparison with oscillation of first-order delay differential equations. Another widely used technique, applicable also in the above-mentioned case, involves the Riccati type transformation which has been used to reduce Equation (1) to a first-order Riccati inequality see (see [2]).

Moaaz et al. [11] improved and extended the Riccati transformation to obtain new oscillatory criteria for the fourth order delay differential equations

$$\left(\pi\left(y\right)\left(z'''\left(y\right)\right)^{\alpha}\right)' + \int_{a}^{m} q\left(y,\xi\right) f\left(z\left(\Phi\left(y,\xi\right)\right)\right) d\sigma\left(\xi\right) = 0, \quad y \geq y_0.$$

Elabbasy et al. [7] studied the equation

$$\left[m\left(y\right)\left(z^{(n-1)}\left(y\right)\right)^{\gamma}\right]' + \sum_{i=1}^{m} q_i\left(y\right) f\left(z\left(\sigma_i\left(y\right)\right)\right) = 0, \quad y \geq y_0.$$

Agarwal et al. [1] and the present authors in [18] used the comparison technique for the fourth order delay differential equation

$$\left[m\left(y\right)\left(z^{(n-1)}\left(y\right)\right)^{\gamma}\right]' + q\left(y\right) z^{\gamma}\left(\sigma\left(y\right)\right) = 0, \quad y \geq y_0,$$

that was compared with the oscillation of certain first order differential equation and under the conditions

$$\int_{y_0}^{\infty} \frac{1}{m^{\frac{1}{\gamma}}(y)} dy = \infty,$$

and

$$\int_{y_0}^{\infty} \frac{1}{m^{\frac{1}{\gamma}}(y)} dy < \infty.$$

However, the authors of this paper used the comparison technique for the fourth order delay differential equation and that was compared with the oscillation of certain second order differential equation.

To the best of our knowledge, there is nothing known about the oscillation of (1) to be oscillatory under the

$$\int_{y_0}^{\infty} \frac{1}{m_i(y)} dy = \infty. \tag{2}$$

Our primary goal is to fill this gap by presenting simple criteria for the oscillation of all solutions of (1). So the main advantage of studying (1) essentially lies in the direct application of the well-known Kiguradze lemma [23] (Lemma 1), which allows one to classify the set of possible nonoscillatory solutions.

In what follows, all occurring functional inequalities are assumed to hold eventually, that is, they are satisfied for all t large enough. As usual and without loss of generality, we can deal only with eventually positive solutions of (1).

2. Main Results

In this section, we state some oscillation criteria for (1). For convenience, we denote

$$\pi_i(y) = \int_{y_1}^{y} \frac{1}{m_i(s)} ds, \quad i = 1, 2, 3, \quad I_2(y) = \int_{y_1}^{y} \frac{1}{m_1(s)} \pi_2(s) ds.$$

$$A_2(y) = \int_{y_1}^{y} \frac{1}{m_2(s)} \pi_3(s) ds, \quad A_3(y) = \int_{y_1}^{y} \frac{1}{m_1(s)} A_2(s) ds.$$

$$\tilde{E}_0 z(y) = z(y), \quad \tilde{E}_i z(y) = m_i (\tilde{E}_{i-1} z(y))', \quad i = 1, 2, 3, \quad \tilde{E}_4 z(y) = (\tilde{E}_3 z(y))'.$$

where y_1 is sufficiently large.

The main step to study Equation (1) is to determine the derivatives sign $\tilde{E}_i z(y)$ according to Kiguradze's lemma [23]

$$\tilde{E}_4 z(y) + q(y) f(z(\sigma(y))) = 0,$$

the set Φ of nonoscillatory solutions can be divided into two parts

$$\Phi = \Phi_1 \cup \Phi_3,$$

say positive solution $z(y)$ satisfies

$$z(y) \in \Phi_1 \iff \tilde{E}_1 z(y) > 0, \quad \tilde{E}_2 z(y) < 0, \quad \tilde{E}_3 z(y) > 0, \quad \tilde{E}_4 z(y) < 0,$$

or

$$z(y) \in \Phi_3 \iff \tilde{E}_1 z(y) > 0, \quad \tilde{E}_2 z(y) > 0, \quad \tilde{E}_3 z(y) > 0, \quad \tilde{E}_4 z(y) < 0.$$

Theorem 1. *Let (2) hold. Assume that $z(y)$ be a positive solution of Equation (1). If*

(i) $z(y) \in \Phi_1$, then $\dfrac{z(y)}{\pi_1(y)}$ is decreasing.

(ii) $z(y) \in \Phi_3$, then $\dfrac{z(y)}{A_3(y)}$ is decreasing and $\tilde{E}_1 z(y) \geq A_2(y) \tilde{E}_3 z(y)$.

Proof. Let $z(y)$ be a positive solution of (1) and $z(y) \in \Phi_1$. It follows from the monotonicity of $\tilde{E}_1 z(y)$ that

$$\begin{aligned} z(y) &> z(y) - z(y_1) \\ &= \int_{y_1}^{y} \frac{1}{m_1(s)} \tilde{E}_1 z(s) ds, \\ &\geq \tilde{E}_1 z(y) \int_{y_1}^{y} \frac{1}{m_1(s)} ds, \\ &\geq \tilde{E}_1 z(y) \pi_1(y) > m_1(y) z'(y) \pi_1(y). \end{aligned}$$

Therefore,

$$\left(\frac{z(y)}{\pi_1(y)} \right)' = \frac{z'(y) \pi_1(y) - z(y) \frac{1}{m_1(y)}}{(\pi_1(y))^2} < 0, \tag{3}$$

case (i) is proved. Now let $z(y) \in \Phi_3$. Since

$$\begin{aligned} \tilde{E}_2 z(y) &= \tilde{E}_2 z(y_1) + \int_{y_1}^{y} \frac{1}{m_3(s)} \tilde{E}_3 z(s) ds \\ &> \tilde{E}_3 z(y) \pi_3(y) \end{aligned}$$

then
$$\left(\frac{\tilde{E}_2 z(y)}{\pi_3(y)}\right)' = \frac{\tilde{E}_2' z(y)\pi_3(y) - \tilde{E}_2 z(y)\frac{1}{m_3(y)}}{(\pi_3(y))^2} < 0. \tag{4}$$

Thus $\dfrac{\tilde{E}_2 z(y)}{\pi_3(y)}$ is decreasing. Moreover,

$$\tilde{E}_1 z(y) = \tilde{E}_1 z(y_1) + \int_{y_1}^{y} \frac{\pi_3(s)}{m_2(s)} \frac{\tilde{E}_2 z(s)}{\pi_3(s)} ds,$$
$$> \frac{\tilde{E}_2 z(y)}{\pi_3(y)} A_2(y).$$

we obtain $\tilde{E}_1 z(y) \geq A_2(y)\tilde{E}_3 z(y)$ and

$$\left(\frac{\tilde{E}_1 z(y)}{A_2(y)}\right)' = \frac{\tilde{E}_1' z(y) A_2(y) - \frac{1}{m_2(y)}\pi_3(y)\tilde{E}_1 z(y)}{(A_2(y))^2} < 0. \tag{5}$$

Thus $\dfrac{\tilde{E}_1 z(y)}{A_2(y)}$ is decreasing. On the other hand,

$$z(y) = z(y_1) + \int_{y_1}^{y} \frac{A_2(s)}{m_1(s)} \frac{\tilde{E}_1 z(s)}{A_2(s)} ds,$$
$$> \frac{\tilde{E}_1 z(y)}{A_2(y)} A_3(y),$$

which implies

$$\left(\frac{z(y)}{A_3(y)}\right)' = \frac{z'(y) A_3(y) - \frac{1}{m_1(y)} A_2(y) z(y)}{(A_3(y))^2} < 0. \tag{6}$$

So that $\dfrac{z(y)}{A_3(y)}$ is decreasing. Theorem is proved. □

Let
$$\delta(y) = \frac{1}{m_1(y)}\left(\int_{y}^{\sigma^{-1}(y)} \frac{1}{m_2(s)} \int_{s}^{\sigma^{-1}(y)} \frac{1}{m_3(v)} dv ds \int_{\sigma^{-1}(y)}^{\infty} kq(s)\, ds\right).$$

Theorem 2. *Let (2) hold. Let $z(y)$ be a positive solution of Equation (1). If*

(i) $z(y) \in \Phi_1$, then $z'(y) \geq \delta(y) z(y)$.
(ii) $z(y) \in \Phi_3$, then $z'(y) \geq \frac{1}{m_1(y)\pi_1(y)} z(y)$.

Proof. Assume that $z(y)$ is a positive solution of (1) and $z(y) \in \Phi_1$. For any $u > y$, we have $\tilde{E}_1 z(y)$ that

$$-\tilde{E}_2 z(y) = \tilde{E}_2 z(u) - \tilde{E}_2 z(y),$$
$$= \int_{y}^{u} \frac{1}{m_3(s)} \tilde{E}_3 z(s)\, ds, \tag{7}$$
$$> \tilde{E}_3 z(u) \int_{y}^{u} \frac{1}{m_3(s)} ds.$$

Multiplying by $\frac{1}{m_2(s)}$ and then integrating from y to u, one gets

$$\tilde{E}_1 z(y) \geq \int_y^u \frac{\tilde{E}_3 z(y)}{m_2(s)} \int_s^u \frac{1}{m_3(v)} dv ds,$$
$$> \tilde{E}_3 z(u) \int_y^u \frac{1}{m_2(s)} \int_s^u \frac{1}{m_3(v)} dv ds. \qquad (8)$$

An integration of (1) from u to ∞, yields

$$\tilde{E}_3 z(u) \geq \int_u^\infty kq(s) z(\sigma(s)) ds,$$
$$\geq z(\sigma(s)) \int_u^\infty kq(s) ds.$$

Combining (7) together with (8) and setting $u = \sigma^{-1}(y)$, we get

$$z'(y) \geq \frac{1}{m_1(y)} \left(\int_y^{\sigma^{-1}(y)} \frac{1}{m_2(s)} \int_s^{\sigma^{-1}(y)} \frac{1}{m_3(v)} dv ds \int_{\sigma^{-1}(y)}^\infty kq(s) ds \right) z(y).$$
$$\geq \delta(y) z(y). \qquad (9)$$

and case (i) is proved. Now let $z(y) \in \Phi_3$. Employing (H_2), the monotonicity of $\tilde{E}_1 z(y)$ and the fact that $\tilde{E}_1 z(y) \to \infty$ as $y \to \infty$, we get

$$z(y) = z(y_1) + \int_{y_1}^y \frac{1}{m_1(s)} \tilde{E}_1 z(s) ds,$$
$$\leq z(y_1) + \tilde{E}_1 z(y) \int_{y_1}^y \frac{1}{m_1(s)} ds,$$
$$= z(y_1) - \tilde{E}_1 z(y) \int_0^{y_1} \frac{1}{m_1(s)} ds + \tilde{E}_1 z(y) \int_0^y \frac{1}{m_1(s)} ds, \qquad (10)$$
$$\leq \tilde{E}_1 z(s) \int_0^y \frac{1}{m_1(s)} ds.$$

The proof is complete now. □

Now, we apply the results of the previous cases to obtain the oscillation conditions of Equation (1). We denote

$$\delta_1(y) = q(y) \frac{\pi_1(\sigma(y))}{\pi_1(y)},$$
$$\delta_2(y) = kq(y) \frac{A_3(\sigma(y))}{A_3(y)}.$$

Theorem 3. *Let (2) hold. Assume there exists a positive continuously differentiable functions $\rho, \vartheta \in C([y_0, \infty))$ such that*

$$\limsup_{y \to \infty} \int_{y_1}^\infty \left[\frac{\rho(v)}{m_2(v)} \int_v^\infty \frac{1}{m_3(u)} \int_u^\infty \delta_1(s) ds du - \frac{m_1(v)(\rho'(v))^2}{4\rho(v)} \right] dv = \infty, \qquad (11)$$

and

$$\limsup_{y \to \infty} \int_{y_1}^\infty \left[\delta_2(v) \vartheta(s) - \frac{m_1(s)(\vartheta'(v))^2}{4\rho(v) A_2(s)} \right] ds = \infty. \qquad (12)$$

Then every solution of Equation (1) is oscillatory.

Proof. Assume that (1) has a nonoscillatory solution $z(y)$. Without loss of generality, we can assume that $z(y)$ is a positive solution of (1). Then either $z(y) \in \Phi_1$ or $z(y) \in \Phi_3$. Now assume that $z(y) \in \Phi_1$. Theorem 1 implies that

$$z(\sigma(y)) \geq \frac{\pi_1(\sigma(y))}{\pi_1(y)} z(y)$$

On the other hand, it follows from Theorem 2 that

$$z'(y) \geq \delta(y) z(y).$$

Setting both estimates into (1), we get

$$\tilde{E}_4 z(y) + \delta_1(y) \leq 0.$$

Integrating from y to ∞ one gets

$$\begin{aligned}
-\tilde{E}_3 z(y) &\geq \int_y^\infty \delta_1(s) z(s)\, ds, \\
&\geq z(y) \int_y^\infty \delta_1(s)\, ds.
\end{aligned} \tag{13}$$

Integrating once more, we have

$$\tilde{E}_2 z(y) + \left(\int_y^\infty \frac{1}{m_3(u)} \int_u^\infty \delta_1(s)\, ds\, du \right) z(y) \leq 0. \tag{14}$$

Define the function $\omega(y)$ by

$$\omega(y) := \rho(y) \frac{\tilde{E}_1 z(y)}{z(y)}, \tag{15}$$

then $\omega(y) > 0$ and

$$\begin{aligned}
\omega'(y) &= \rho'(y) \frac{\tilde{E}_1 z(y)}{z(y)} + \rho(y) \frac{\tilde{E}_2 z(y)}{m_2(y) z(y)} - \rho(y) \frac{\tilde{E}_1 z(y) z'(y)}{z^2(y)} \\
&\leq -\frac{\rho(y)}{m_2(y) z(y)} \int_y^\infty \frac{1}{m_3(u)} \int_u^\infty \delta_1(s)\, ds\, du + \frac{\rho'(y)}{\rho(y)} \omega(y) - \frac{\omega^2(y)}{m_1(y) \rho(y)} \\
&\leq -\frac{\rho(y)}{m_2(y) z(y)} \int_y^\infty \frac{1}{m_3(u)} \int_u^\infty \delta_1(s)\, ds\, du + \frac{m_1(y) (\rho'(y))^2}{4\rho(y)}.
\end{aligned} \tag{16}$$

Integration of the previous inequality yields

$$\int_{y_1}^y \left[\frac{\rho(v)}{m_1(v)} \int_v^\infty \frac{1}{m_3(u)} \int_u^\infty \delta_1(s)\, ds\, du - \frac{m_1(v)(\rho'(v))^2}{4\rho(v)} \right] dv \leq \omega(y_1),$$

this contradicts with (11) as $y \to \infty$. Now assume that $z(y) \in \Phi_3$. Theorems 1 and 2 guarantee that

$$z(\sigma(y)) \geq \frac{A_3(\sigma(y))}{A_3(y)} z(y), \quad z'(y) \geq \frac{1}{m_1(y)\pi_1(y)} z(y), \quad \tilde{E}_1 z(y) \geq A_2(y) \tilde{E}_3 z(y),$$

what in view of (1) provides

$$\tilde{E}_4 z(y) + \delta_2(y) \leq 0.$$

Now define $\psi(y)$ by

$$\psi(y) := \vartheta(y) \frac{\tilde{E}_3 z(y)}{z(y)}, \tag{17}$$

then $\psi(y) > 0$ and

$$\begin{aligned}
\psi'(y) &= \vartheta'(y)\frac{\bar{E}_3 z(y)}{z(y)} + \vartheta(y)\frac{\bar{E}_4 z(y)}{z(y)} - \vartheta(y)\frac{\bar{E}_3 z(y) z'(y)}{z^2(y)} \\
&\leq -\vartheta(y)\delta_2(y) + \frac{\vartheta'(y)}{\vartheta(y)}\psi(y) - \frac{A_2(y)\psi^2(y)}{m_1(y)\vartheta(y)} \\
&\leq -\vartheta(y)\delta_2(y) + \frac{m_1(y)(\vartheta'(y))^2}{4\vartheta(y) A_2}.
\end{aligned} \qquad (18)$$

Integrating from y_1 to y and letting $y \to \infty$, we get

$$\int_{y_1}^{\infty}\left[\delta_2(v)\vartheta(s) - \frac{m_1(s)(\vartheta'(v))^2}{4\rho(v) A_2(s)}\right]ds \leq \psi(y_1),$$

which contradicts with (12) and the proof is complete. □

Corollary 1. *Let (2) hold and*

$$\limsup_{y\to\infty}\int_{y_1}^{\infty}\left[\frac{\pi_1(v)}{m_2(v)}\int_v^{\infty}\frac{1}{m_3(u)}\int_u^{\infty}\delta_1(s)\,ds\,du - \frac{1}{4m_1(v)\pi_1(v)}\right]dv = \infty, \qquad (19)$$

$$\limsup_{y\to\infty}\int_{y_1}^{\infty}\left[kq(s) A_3(\sigma(s)) - \frac{A_2(s)}{4m_1(s) A_3(s)}\right]ds = \infty. \qquad (20)$$

Then every solution of Equation (1) is oscillatory.

Now, we use the comparison method to obtain other oscillation results. It is well known (see [10]) that the differential equation

$$[a(y)(z'(y))]' + q(y) z(\sigma(y)) = 0, \quad y \geq y_0, \qquad (21)$$

where $a, q \in C[y_0, \infty)$, $a(y), q(y) > 0$, is nonoscillatory if and only if there exists a number $y \geq y_0$, and a function $v \in C^1[y, \infty)$, satisfying the inequality

$$v'(y) + \alpha a^{-1}(y) v^2(y) + q(y) \leq 0, \quad \text{on } [y, \infty).$$

Lemma 1 (see [10]). *Let*

$$\int_{y_0}^{\infty}\frac{1}{a(s)}ds = \infty$$

holds, then the condition

$$\liminf_{y\to\infty}\left(\int_{y_0}^{\infty}\frac{1}{a(s)}ds\right)\int_y^{\infty}q(s)\,ds > \frac{1}{4}.$$

Theorem 4. *Let (2) hold. Assume that the equation*

$$[m_1(y) z'(y)]' + \left(\frac{1}{m_2(y)}\int_y^{\infty}\frac{1}{m_3(u)}\int_u^{\infty}\delta_1(s)\,ds\,du\right) z(y) = 0, \qquad (22)$$

and

$$\left(\frac{m_1(y)}{\pi_3(y)} z'(y)\right)' + \delta_2(y) z(y) = 0, \qquad (23)$$

are oscillatory, then every solution of (1) is oscillatory.

Proof. Proceeding as in proof of the Theorem 3. We get (16). If we set $\rho(y) = 1$ in (16), then we obtain

$$\omega'(y) + \frac{1}{m_1(y)}\omega^2(y) - \frac{1}{m_2(y)}\int_y^\infty \frac{1}{m_3(u)}\int_u^\infty \delta_1(s)\,ds\,du \leq 0.$$

Thus, we can see that Equation (22) is nonoscillatory for every constant $\lambda_1 \in (0,1)$, which is a contradiction. If we now set $\vartheta(y) = 1$ in (18), then we find

$$\psi'(y) + \frac{A_2(y)}{m_1(y)}\psi^2(y) + \delta_2(y) \leq 0.$$

Hence, Equation (23) is nonoscillatory, which is a contradiction.
Theorem 4 is proved. □

In view of Lemma 1, oscillation criteria for (1) of Hille–Nehari-type are easily acquired. Please note that

Corollary 2. *Assume that*

$$\liminf_{y\to\infty} \pi_1(y) \int_y^\infty \frac{1}{m_2(v)}\int_v^\infty \frac{1}{m_3(u)}\int_u^\infty \delta_1(s)\,ds\,du > \frac{1}{4},$$

$$\liminf_{y\to\infty} \left(\int_{y_0}^y \frac{A_2(s)}{m_1(s)}ds\right) \int_y^\infty \delta_2(s)\,ds > \frac{1}{4}.$$

Then every solution of (1) is oscillatory.

3. Example

In this section, we give the following example to illustrate our main results.

Example 1. *Let us consider the fourth-order differential equation of type*

$$\left(y^{1\backslash 2}\left(y^{1\backslash 2}z'(y)\right)''\right)' + \frac{1}{y^3}z(\beta y) = 0,\ y \geq 1, \tag{24}$$

where $0 < \beta < 1$ is a constant. Let

$$m_3(y) = y^{1\backslash 2},\ m_2(y) = y^{1\backslash 2},\ m_1(y) = 1 > 0,\ q(y) = \frac{1}{y^3},\ \sigma(y) = \beta y,$$

and

$$\pi_i(s) := \int_{y_0}^\infty \frac{1}{m_i(s)}ds = \infty.$$

If we now set $k = 1$, It is easy to see that all conditions of Corollary 1 are satisfied.

$$\begin{aligned}
A_3(\sigma(s)) &= \int_{\sigma_1(s)}^{\sigma(s)} \frac{1}{m_1(\sigma(s))} A_2(\sigma(s))\,ds \\
&= \int_{\sigma_1(s)}^{\sigma(s)} \left(\int_{\sigma_1(s)}^{\sigma(s)} \frac{1}{m_2(\sigma(s))}\pi_3(\sigma(s))\,ds\right)ds \\
&= \int_{\sigma_1(s)}^{\sigma(s)} \left(\int_{\sigma_1(s)}^{\sigma(s)} \frac{1}{(\beta s)^{1/2}}\left(\int_{\sigma_1(s)}^\infty \frac{1}{(\beta s)^{1/2}}ds\right)ds\right)ds
\end{aligned}$$

$$A_2(s) = \int_{s_1}^{s} \frac{1}{m_2(s)} \pi_3(s)\, ds$$
$$= \int_{s_1}^{s} \frac{1}{s^{1/2}} \left(\int_{s_1}^{\infty} \frac{1}{s^{1/2}} ds \right) ds$$

now

$$\limsup_{y \to \infty} \int_{y_1}^{\infty} \left[kq(s) A_3(\sigma(s)) - \frac{A_2(s)}{4m_1(s) A_3(s)} \right] ds$$

$$\limsup_{y \to \infty} \int_{y_1}^{\infty} \left[\frac{1}{s^3} \int_{\sigma_1(s)}^{\sigma(s)} \left(\int_{\sigma_1(s)}^{\sigma(s)} \frac{1}{(\beta s)^{1/2}} \left(\int_{\sigma_1(s)}^{\infty} \frac{1}{(\beta s)^{1/2}} ds \right) ds \right) ds \right] ds$$
$$= \infty$$

and

$$\limsup_{y \to \infty} \int_{y_1}^{\infty} \left[\frac{\pi_1(v)}{m_2(v)} \int_{v}^{\infty} \frac{1}{m_3(u)} \int_{u}^{\infty} \delta_1(s)\, ds du - \frac{1}{4m_1(v)\, \pi_1(v)} \right] dv = \infty.$$

Hence, by Corollary 1, every solution of Equation (25) is oscillatory.

Example 2. *Consider a differential equation*

$$\left(y \left(y \left(y z'(y) \right)' \right)' \right)' + y z(y) = 0, \ y \geq 1, \tag{25}$$

We see

$$m_3(y) = m_2(y) = m_1(y) = y > 0,\ q(y) = y,\ \sigma(y) = y,$$

and

$$\pi_i(s) := \int_{y_0}^{\infty} \frac{1}{m_i(s)} ds = \infty.$$

If we now set $k = 1$, It is easy to see that all conditions of Corollary 1 are satisfied.

$$\limsup_{y \to \infty} \int_{y_1}^{\infty} \left[\frac{\pi_1(v)}{m_2(v)} \int_{v}^{\infty} \frac{1}{m_3(u)} \int_{u}^{\infty} \delta_1(s)\, ds du - \frac{1}{4m_1(v)\, \pi_1(v)} \right] dv = \infty,$$

$$\limsup_{y \to \infty} \int_{y_1}^{\infty} \left[kq(s) A_3(\sigma(s)) - \frac{A_2(s)}{4m_1(s) A_3(s)} \right] ds = \infty.$$

Hence, by Corollary 1, every solution of Equation (25) is oscillatory.

4. Conclusions

The results of this paper are presented in a form which is essentially new and of high degree of generality. To the best of our knowledge, there is nothing known about the oscillation of (1) under the assumption (2), our primary goal is to fill this gap by presenting simple criteria for the oscillation of all solutions of (1) by using the generalized Riccati transformations and comparison technique, so the main advantage of studying (1) essentially lies in the direct application of the well-known Kiguradze lemma [23] (Lemma 1). Further, we can consider the case of $\sigma(y) \geq y$ in the future work.

Author Contributions: The authors claim to have contributed equally and significantly in this paper. All authors read and approved the final manuscript.

Funding: The authors declare that they have not received funds from any institution.

Acknowledgments: The authors thank the referee for carefully reading the manuscript and suggesting very useful comments which improve the content of the paper.

Conflicts of Interest: The authors declare no conflict of interest.

References

1. Agarwal, R.; Grace, S.; O'Regan, D. Oscillation criteria for certain nth order differential equations with deviating arguments. *J. Math. Appl. Anal.* **2001**, *262*, 601–622. [CrossRef]
2. Baculikova, B.; Dzurina, J.; Graef, J.R. On the oscillation of higher-order delay differential equations. *Math. Slovaca* **2012**, *187*, 387–400. [CrossRef]
3. Bazighifan, O. *Oscillation Criteria For Nonlinear Delay Differential Equation*; Lambert Academic Publishing: Saarbrucken, Germany, 2017.
4. Bazighifan, O. Oscillatory behavior of higher-order delay differential equations. *Gen. Lett. Math.* **2017**, *2*, 105–110.
5. Bazighifan, O.; Elabbasy, E.M.; Moaaz, O. Oscillation of higher-order differential equations with distributed delay. *J. Inequal. Appl.* **2019**, *2019*, 55. [CrossRef]
6. Bazighifan, O. On Oscillatory and Asymptotic Behavior of Fourth Order Nonlinear Neutral Delay Differential Equations. *Int. J. Mod. Math. Sci.* **2019**, *17*, 21–30.
7. Elabbasy, E.M.; Moaaz, O.; Bazighifan, O. Oscillation Solution for Higher-Order Delay Differential Equations. *J. King Abdulaziz Univ.* **2017**, *29*, 45–52.
8. Elabbasy, E.M.; Moaaz, O.; Bazighifan, O. Oscillation of Fourth-Order Advanced Differential Equations. *J. Mod. Sci. Eng.* **2017**, *3*, 64–71.
9. Elabbasy, E.M.; Moaaz, O.; Bazighifan, O. Oscillation Criteria for Fourth-Order Nonlinear Differential Equations. *Int. J. Mod. Math. Sci.* **2017**, *15*, 50–57.
10. Agarwal, R.; Grace, S.; O'Regan, D. *Oscillation Theory for Difference and Functional Differential Equations*; Kluwer Acad. Publ.: Dordrecht, The Netherlands, 2000.
11. Moaaz, O.; Elabbasy, E.M.; Bazighifan, O. On the asymptotic behavior of fourth-order functional differential equations. *Adv. Differ. Equ.* **2017**, *2017*, 261. [CrossRef]
12. Gyori, I.; Ladas, G. *Oscillation Theory of Delay Differential Equations with Applications*; Clarendon Press: Oxford, UK, 1991.
13. Ladde, G.; Lakshmikantham, V.; Zhang, B. *Oscillation Theory of Differential Equations with Deviating Arguments*; Marcel Dekker: New York, NY, USA, 1987.
14. Grace, S.; Agarwal, R.; Graef, J. Oscillation theorems for fourth order functional differential equations. *J. Appl. Math. Comput.* **2009**, *30*, 75–88. [CrossRef]
15. Li, T.; Baculikova, B.; Dzurina, J.; Zhang, C. Oscillation of fourth order neutral differential equations with p-Laplacian like operators. *Bound. Value Probl.* **2014**, *56*, 41–58. [CrossRef]
16. Moaaz, O. *Oscillation Properties of Some Differential Equations*; Lambert Academic Publishing: Saarbrucken, Germany, 2017.
17. Philos, C. On the existence of nonoscillatory solutions tending to zero at ∞ for differential equations with positive delay. *Arch. Math.* **1981**, *36*, 168–178. [CrossRef]
18. Zhang, C.; Agarwal, R.; Bohner, M.; Li, T. New results for oscillatory behavior of even-order half-linear delay differential equations. *Appl. Math. Lett.* **2013**, *26*, 179–183. [CrossRef]
19. Agarwal, R.; Grace, S.; Manojlovic, J. Oscillation criteria for certain fourth order nonlinear functional differential equations. *Math. Comput. Model.* **2006**, *44*, 163–187. [CrossRef]
20. Assante, D.; Cesarano, C.; Foranaro, C.; Vazquez, L. Higher Order and Fractional Diffusive Equations. *J. Eng. Sci. Technol. Rev.* **2015**, *8*, 202–204. [CrossRef]
21. Cesarano, C.; Foranaro, C.; Vazquez, L. Operational results in bi-orthogonal Hermite functions. *Acta Math. Univ. Comen.* **2016**, *85*, 43–68.
22. Cesarano, C.; Cennamo, G.; Placdi, L. Humbert Polynomials and Functions in Terms of Hermite Polynomials Towards Applications to Wave Propagation. *Wesas Trans. Math.* **2014**, *13*, 595–602.
23. Kiguradze, I.; Chanturia, T. *Asymptotic Properties of Solutions of Nonautonomous Ordinary Differential Equations*; Kluwer Acad. Publ.: Drodrcht, The Netherlands, 1993.

 © 2019 by the authors. Licensee MDPI, Basel, Switzerland. This article is an open access article distributed under the terms and conditions of the Creative Commons Attribution (CC BY) license (http://creativecommons.org/licenses/by/4.0/).

Article

The Third and Fourth Kind Pseudo-Chebyshev Polynomials of Half-Integer Degree

Clemente Cesarano [1,*], Sandra Pinelas [2] and Paolo Emilio Ricci [1]

[1] Section of Mathematics, International Telematic University Uninettuno, Corso Vittorio Emanuele II, 39, 00186 Roma, Italy; paoloemilioricci@gmail.com
[2] Academia Militar, Departamento de Ciências Exactas e Engenharia, Av. Conde Castro Guimarães, 2720-113 Amadora, Portugal; sandra.pinelas@gmail.com
* Correspondence: c.cesarano@uninettunouniversity.net

Received: 30 January 2019; Accepted: 19 February 2019; Published: 20 February 2019

Abstract: New sets of orthogonal functions, which correspond to the first, second, third, and fourth kind Chebyshev polynomials with half-integer indexes, have been recently introduced. In this article, links of these new sets of irrational functions to the third and fourth kind Chebyshev polynomials are highlighted and their connections with the classical Chebyshev polynomials are shown.

Keywords: Chebyshev polynomials; pseudo-Chebyshev polynomials; recurrence relations; differential equations; composition properties; orthogonality properties

1. Introduction

In the second half of the XIX Century, Pafnuty Lvovich Chebyshev introduced two sets of polynomials, presently known as the first and second kind Chebyshev polynomials, which are actually a polynomial version of the circular sine and cosine functions. These polynomials have proved to be of fundamental importance in many questions of an applicative nature (see the classical book by T. Rivlin [1]). In fact, the roots of the first kind polynomials—the so called Chebyshev nodes—appear in approximation theory, since, by using these nodes, the relevant Gaussian quadrature rule realizes the highest possible degree of precision. Moreover, the resulting interpolation polynomial minimizes the Runge phenomenon. Furthermore, by expanding a continuous function in terms of first kind Chebyshev polynomials, the best approximation, with respect to the maximum norm, can be obtained. The second kind Chebyshev polynomials appear in the computation of powers of 2×2 non-singular matrices [2]. For the same problem, in the case of powers of higher order matrices, an extension of these polynomials have been also introduced (see, e.g., [3,4]).

It is also useful to notice that Chebyshev polynomials represent an important tool in deriving integral representations [5,6], and that they can be generalized by using the properties and formalism of the Hermite polynomials [7]; for instance, by introducing multi-variable polynomials recognized as belonging to the Chebyshev family [8–10].

An excellent book on this subject is [11]. The importance of the Chebyshev polynomials in applications has been highlighted, in [12]. In a recent paper, the Chebyshev polynomials of the first and second kind have been shown to represent the real and imaginary part, respectively, of the complex Appell polynomials [13].

In a recent article [14], new sets of functions related to the classical Chebyshev polynomials have been introduced, in connections with the complex version of the Bernoulli spiral. Actually, the real and imaginary part of the Bernoulli spirals define the Rodhonea (or Grandi) curves of fractional index, which often appear in natural shapes [15]. This allows us to define two sets of functions corresponding to the first and second kind Chebyshev polynomials with fractional degree, called pseudo-Chebyshev polynomials (or pseudo-Chebyshev functions), as they are irrational functions. It was shown that, in the case of half-integer degree, the relevant pseudo-Chebyshev polynomials are orthogonal in the interval $(-1,1)$, with respect to the same weights of the Chebyshev polynomials of the same type.

In this article, by using the results of [16], we show the connections of the third and fourth kind pseudo-Chebyshev polynomials with the classical Chebyshev polynomials.

2. Definitions of Pseudo-Chebyshev Functions

The following polynomials $T_k(x)$, $U_k(x)$, $V_k(x)$, and $W_k(x)$ denote, respectively, the first, second, third, and fourth kind classical Chebyshev polynomials.

We have, by definition, for any integer k:

$$T_{k+\frac{1}{2}}(x) = \cos\left((k+\tfrac{1}{2})\arccos(x)\right),$$

$$\sqrt{1-x^2}\, U_{k-\frac{1}{2}}(x) = \sin\left((k+\tfrac{1}{2})\arccos(x)\right), \tag{1}$$

$$\sqrt{1-x^2}\, V_{k+\frac{1}{2}}(x) = \cos\left((k+\tfrac{1}{2})\arccos(x)\right), \text{ and}$$

$$W_{k+\frac{1}{2}}(x) = \sin\left((k+\tfrac{1}{2})\arccos(x)\right).$$

Note that definition (1) holds even for negative integer—that is, for $k+1/2 < 0$—according to the parity properties of the trigonometric functions.

The first, second, third, and fourth kind pseudo-Chebyshev functions are represented, in terms of the third and fourth kind Chebyshev polynomials, as follows:

$$T_{k+\frac{1}{2}}(x) = \sqrt{\tfrac{1+x}{2}}\, V_k(x),$$

$$\sqrt{1-x^2}\, U_{k-\frac{1}{2}}(x) = \sqrt{\tfrac{1}{2(1+x)}}\, W_k(x), \tag{2}$$

$$\sqrt{1-x^2}\, V_{k+\frac{1}{2}}(x) = \sqrt{\tfrac{1}{2(1-x)}}\, V_k(x), \text{ and}$$

$$W_{k+\frac{1}{2}}(x) = \sqrt{\tfrac{1-x}{2}}\, W_k(x).$$

3. Properties of the First and Second Kind Pseudo-Chebyshev Functions

3.1. The First Kind Pseudo-Chebyshev $T_{k+1/2}$

In this section, we recall the main properties of the first kind pseudo-Chebyshev functions (their first few graphs are shown in Figure 1).

Recurrence relation

$$\begin{cases} T_{k+\frac{1}{2}}(x) = 2x\, T_{k-\frac{1}{2}}(x) - T_{k-\frac{3}{2}}(x), \\ T_{\pm\frac{1}{2}}(x) = \sqrt{\tfrac{1+x}{2}}. \end{cases} \tag{3}$$

Differential equation

$$(1-x^2)y'' - xy' + \left(k+\tfrac{1}{2}\right)^2 y = 0. \tag{4}$$

Orthogonality property

$$\int_{-1}^{1} T_{h+\frac{1}{2}}(x)\, T_{k+\frac{1}{2}}(x)\, \frac{1}{\sqrt{1-x^2}}\, dx = 0, \qquad (h \neq k), \tag{5}$$

where h, k are integer numbers such that $h + k = 2n$, $n = 1, 2, 3, \ldots$,

$$\int_{-1}^{1} T_{k+\frac{1}{2}}^{2}(x)\, \frac{1}{\sqrt{1-x^2}}\, dx = \frac{\pi}{2}. \tag{6}$$

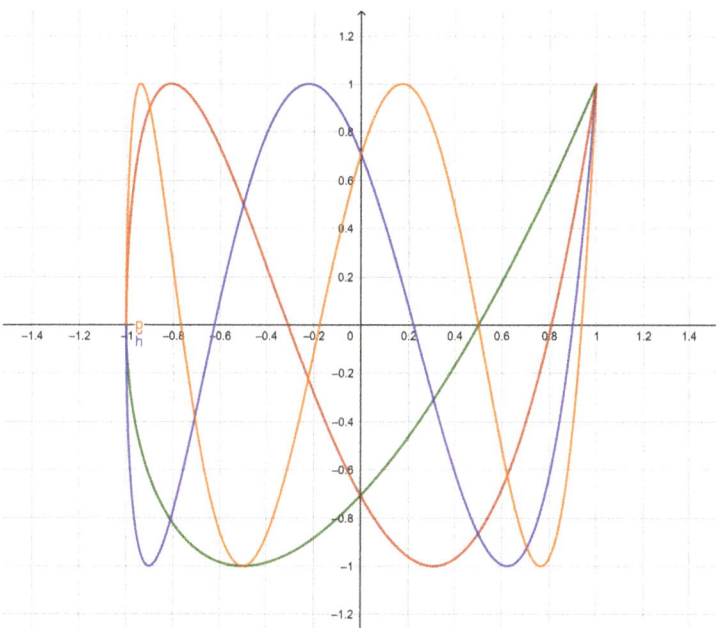

Figure 1. Pseudo-Chebyshev polynomials of the first kind, $T_{k+1/2}(x)$, $k = 1, 2, 3, 4$, where k is: 1, Green; 2, red; 3, blue; and 4, orange.

3.2. The Second Kind Pseudo-Chebyshev $U_{k+1/2}$

In this section we recall the main properties of the second kind pseudo-Chebyshev functions (their first few graphs are shown in Figure 2).

Recurrence relation

$$\begin{cases} U_{k+\frac{1}{2}}(x) = 2x\, U_{k-\frac{1}{2}}(x) - U_{k-\frac{3}{2}}(x), \\ U_{-\frac{1}{2}}(x) = \frac{1}{\sqrt{2(1+x)}}, \quad U_{\frac{1}{2}}(x) = \frac{2x+1}{\sqrt{2(1+x)}}. \end{cases} \tag{7}$$

Differential equation

$$(1 - x^2)\, y'' - 3x\, y' + \left(k - \tfrac{1}{2}\right)\left(k + \tfrac{3}{2}\right) y = 0. \tag{8}$$

Orthogonality property

$$\int_{-1}^{1} U_{h+\frac{1}{2}}(x)\, U_{k+\frac{1}{2}}(x) \sqrt{1-x^2}\, dx = 0, \qquad (h \ne k), \tag{9}$$

where h,k are integer numbers such that $h+k = 2n$, $n = 1,2,3,\ldots$,

$$\int_{-1}^{1} U_{k+\frac{1}{2}}^2(x) \sqrt{1-x^2}\, dx = \frac{\pi}{2}. \tag{10}$$

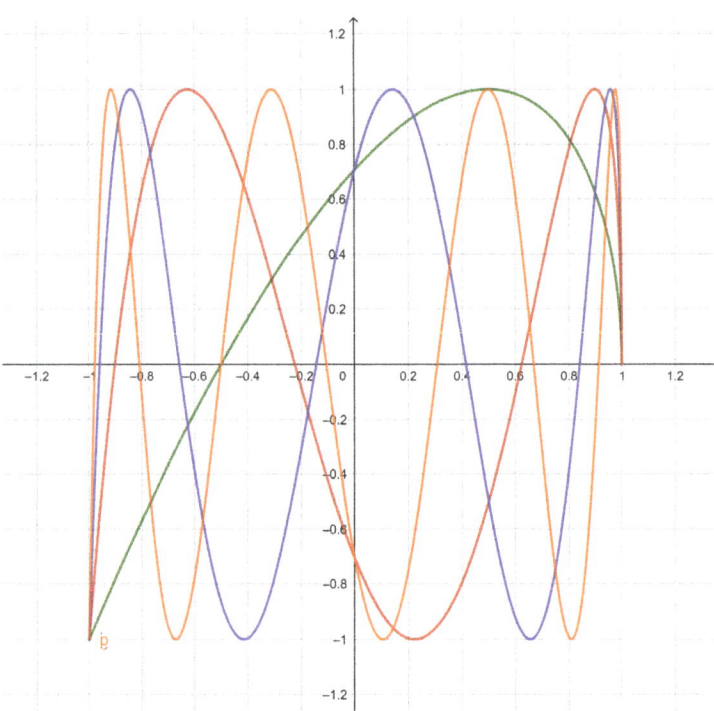

Figure 2. Pseudo-Chebyshev polynomials of the second kind, $U_{k+1/2}(x)$, $k = 1,2,3,4$, where k is: 1, Green; 2, red; 3, blue; and 4, orange.

4. The Third and Fourth Kind Pseudo-Chebyshev Functions

The third and fourth kind Chebyshev polynomials have been also introduced, and studied by several authors (see [16–18]), because they can be applied in particular quadrature rules, where the singularity of the considered function appears at only one of the extrema (+1 or −1) of the integration interval (see [11]). Moreover, in a recent article, they have been used in the framework of solving high odd-order boundary value problems [17].

In what follows, we use the excellent survey by K. Aghigh, M. Masjed-Jamei, and M. Dehghan [16], which permits us to derive, in a straightforward way, the links among the pseudo-Chebyshev functions and the third and fourth kind Chebyshev polynomials.

In Figures 3 and 4, graphs of the first few third and fourth kind pseudo-Chebyshev functions are shown.

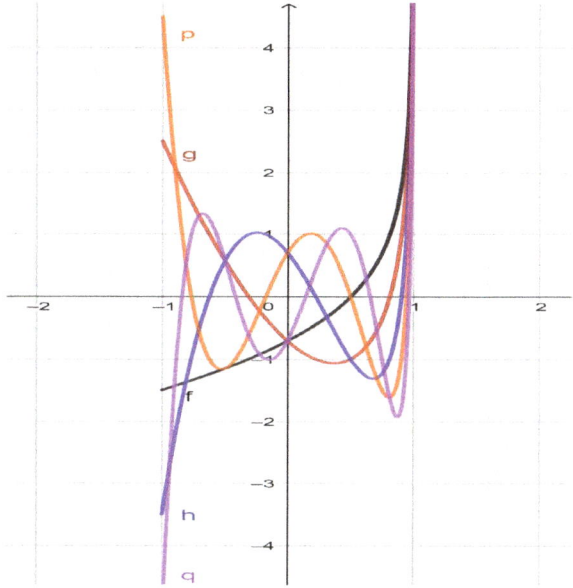

Figure 3. Pseudo-Chebyshev polynomials of the third kind, $V_{k+1/2}(x)$, $k = 1, 2, 3, 4, 5$, where k is: 1, Grey; 2, red; 3, blue; 4, orange; and 5, violet.

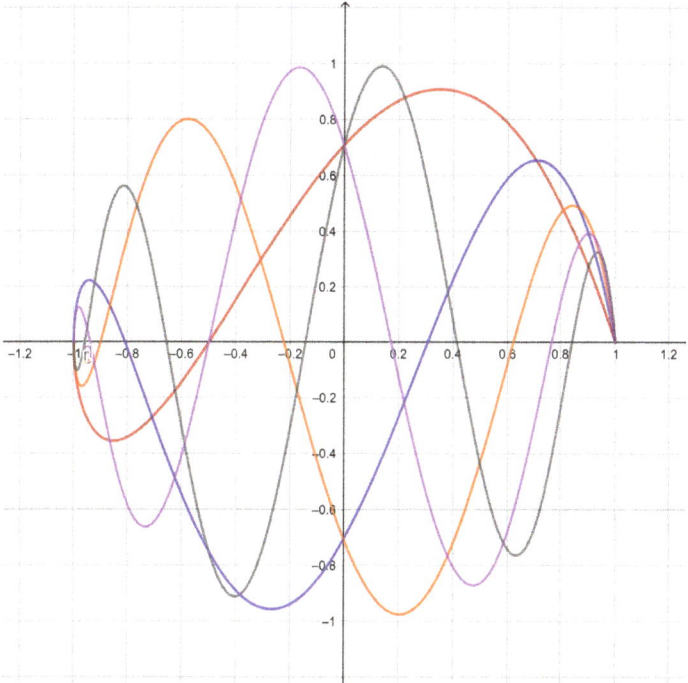

Figure 4. Pseudo-Chebyshev polynomials of the fourths kind, $W_{k+1/2}(x)$, $k = 1, 2, 3, 4, 5$, where k is: 1, Red; 2, blue; 3, orange; 4, violet; and 5, grey.

4.1. The Third Kind Pseudo-Chebyshev $V_{k+1/2}$

Recurrence relation

$$\begin{cases} V_{k+\frac{1}{2}}(x) = 2x\, V_{k-\frac{1}{2}}(x) - V_{k-\frac{3}{2}}(x), \\ V_{\pm\frac{1}{2}}(x) = \dfrac{1}{\sqrt{2(1-x)}}. \end{cases} \qquad (11)$$

Differential equation

Theorem 1. *The third kind pseudo-Chebyshev functions $V_{k+1/2}(x)$ satisfy the differential equation:*

$$(1 - x^2)\, y'' - 3x\, y' + \left(k - \tfrac{1}{2}\right)\left(k + \tfrac{3}{2}\right) y = 0, \qquad (12)$$

so that the second and third kind pseudo-Chebyshev functions are solutions of the same differential equation.

Proof. Note that, from definition (1):

$$D_x V_{k+\frac{1}{2}}(x) = \frac{x}{1-x^2} V_{k+\frac{1}{2}}(x) + \left(k + \tfrac{1}{2}\right) \frac{1}{1-x^2} W_{k+\frac{1}{2}}(x),$$

$$D_x^2 V_{k+\frac{1}{2}}(x) = -\left[\left(k + \tfrac{1}{2}\right)^2 - 1\right] \frac{1}{1-x^2} V_{k+\frac{1}{2}}(x) + \frac{3x^2}{(1-x^2)^2} V_{k+\frac{1}{2}}(x) +$$

$$+ 3\left(k + \tfrac{1}{2}\right) \frac{x}{(1-x^2)^2} W_{k+\frac{1}{2}}(x),$$

$$D_x^2 V_{k+\frac{1}{2}}(x) - \frac{3x}{1-x^2} D_x V_{k+\frac{1}{2}}(x) = -\left[\left(k - \tfrac{1}{2}\right)\left(k + \tfrac{3}{2}\right)\right] V_{k+\frac{1}{2}}(x),$$

so that Equation (12) follows. □

Orthogonality property

$$\int_{-1}^{1} V_{h+\frac{1}{2}}(x)\, V_{k+\frac{1}{2}}(x)\, \sqrt{1-x^2}\, dx = 0, \qquad (h \neq k), \qquad (13)$$

where h,k are integer numbers such that $h + k = 2n$, $n = 1, 2, 3, \ldots$,

$$\int_{-1}^{1} V_{k+\frac{1}{2}}^2(x)\, \sqrt{1-x^2}\, dx = \frac{\pi}{2}. \qquad (14)$$

4.2. The Fourth Kind Pseudo-Chebyshev $W_{k+1/2}$

Recurrence relation

$$\begin{cases} W_{k+\frac{1}{2}}(x) = 2x\, W_{k-\frac{1}{2}}(x) - W_{k-\frac{3}{2}}(x), \\ W_{\pm\frac{1}{2}}(x) = \pm\sqrt{\dfrac{1-x}{2}}. \end{cases} \qquad (15)$$

Differential equation

Theorem 2. *The fourth kind pseudo-Chebyshev functions $W_{k+1/2}(x)$ satisfy the differential equation:*

$$(1 - x^2)\, y'' + x\, y' + \left(k + \tfrac{1}{2}\right)^2 (1 - x^2)\, y = 0. \qquad (16)$$

Proof. Note that

$$D_x W_{k+\frac{1}{2}}(x) = -\left(k+\tfrac{1}{2}\right)(1-x^2)^{-1/2} T_{k+\frac{1}{2}}(x),$$

$$D_x^2 W_{k+\frac{1}{2}}(x) = -\left(k+\tfrac{1}{2}\right)^2 W_{k+\frac{1}{2}}(x) - \left(k+\tfrac{1}{2}\right) x (1-x^2)^{-3/2} T_{k+\frac{1}{2}}(x) =$$

$$= -\left(k+\tfrac{1}{2}\right)^2 W_{k+\frac{1}{2}}(x) - x(1-x^2)^{-1} D_x W_{k+\frac{1}{2}}(x),$$

so that Equation (16) follows. □

Orthogonality property

$$\int_{-1}^{1} W_{h+\frac{1}{2}}(x) W_{k+\frac{1}{2}}(x) \frac{1}{\sqrt{1-x^2}} dx = 0, \qquad (h \neq k), \tag{17}$$

where h, k are integer numbers such that $h + k = 2n$, $n = 1, 2, 3, \ldots$,

$$\int_{-1}^{1} W_{k+\frac{1}{2}}^2(x) \frac{1}{\sqrt{1-x^2}} dx = \frac{\pi}{2}. \tag{18}$$

5. Further Properties of the Pseudo-Chebyshev Functions

5.1. Generating Functions

Theorem 3. *The generating functions of the pseudo-Chebyshev functions are given by:*

$$\sum_{k=0}^{\infty} T_{k+\frac{1}{2}}(x) t^k = \sqrt{\tfrac{1+x}{2}} \, \frac{1-t}{1-2tx+t^2},$$

$$\sum_{k=0}^{\infty} U_{k-\frac{1}{2}}(x) t^k = \sqrt{\tfrac{1}{2(1+x)}} \, \frac{1+t}{1-2tx+t^2},$$

$$\sum_{k=0}^{\infty} V_{k+\frac{1}{2}}(x) t^k = \sqrt{\tfrac{1}{2(1-x)}} \, \frac{1-t}{1-2tx+t^2}, \quad \text{and} \tag{19}$$

$$\sum_{k=0}^{\infty} W_{k+\frac{1}{2}}(x) t^k = \sqrt{\tfrac{1-x}{2}} \, \frac{1+t}{1-2tx+t^2}.$$

Proof. Equations (19) follow from Definitions (2) by using the generating functions of the third and fourth Chebyshev polynomials, which are given below (see [16]):

$$\sum_{k=0}^{\infty} V_k(x) t^k = \frac{1-t}{1-2tx+t^2} \quad \text{and}$$

$$\sum_{k=0}^{\infty} W_k(x) t^k = \frac{1+t}{1-2tx+t^2}. \tag{20}$$

□

5.2. Explicit Forms

Theorem 4. *The explicit forms of the pseudo-Chebyshev functions are given by:*

$$T_{k+\frac{1}{2}}(x) = \sqrt{\frac{1+x}{2}} \sum_{h=0}^{k} (-1)^h \binom{2k+1}{2h} \left(\frac{1-x}{2}\right)^h \left(\frac{1+x}{2}\right)^{k-h},$$

$$U_{k-\frac{1}{2}}(x) = \sqrt{\frac{1}{2(1+x)}} \sum_{h=0}^{k} (-1)^h \binom{2k+1}{2h+1} \left(\frac{1-x}{2}\right)^h \left(\frac{1+x}{2}\right)^{k-h},$$

$$V_{k+\frac{1}{2}}(x) = \sqrt{\frac{1}{2(1-x)}} \sum_{h=0}^{k} (-1)^h \binom{2k+1}{2h} \left(\frac{1-x}{2}\right)^h \left(\frac{1+x}{2}\right)^{k-h}, \text{ and}$$

$$W_{k+\frac{1}{2}}(x) = \sqrt{\frac{1-x}{2}} \sum_{h=0}^{k} (-1)^h \binom{2k+1}{2h+1} \left(\frac{1-x}{2}\right)^h \left(\frac{1+x}{2}\right)^{k-h}.$$

(21)

Proof. Recalling that

$$\cos\left(\tfrac{1}{2} \arccos(x)\right) = \sqrt{\tfrac{1+x}{2}} \quad \text{and} \quad \sin\left(\tfrac{1}{2} \arccos(x)\right) = \sqrt{\tfrac{1-x}{2}}, \tag{22}$$

we find:

$$\left[\cos\left(\tfrac{1}{2} \arccos(x)\right) + i \sin\left(\tfrac{1}{2} \arccos(x)\right)\right]^{2k+1} = \left(\sqrt{\tfrac{1+x}{2}} + i\sqrt{\tfrac{1-x}{2}}\right)^{2k+1}, \tag{23}$$

so that, by the binomial theorem, we find (see [16]):

$$\left(\sqrt{\tfrac{1+x}{2}} + i\sqrt{\tfrac{1-x}{2}}\right)^{2k+1} = \sqrt{\tfrac{1+x}{2}} \sum_{h=0}^{k} (-1)^h \binom{2k+1}{2h} \left(\tfrac{1-x}{2}\right)^h \left(\tfrac{1+x}{2}\right)^{k-h}$$

$$+ i\sqrt{\tfrac{1-x}{2}} \sum_{h=0}^{k} (-1)^h \binom{2k+1}{2h+1} \left(\tfrac{1-x}{2}\right)^h \left(\tfrac{1+x}{2}\right)^{k-h}.$$

(24)

Therefore, recalling Definitions (2), Equation (21) follows. □

5.3. Location of Zeros

By Equation (1), the zeros of the pseudo-Chebyshev functions $T_{k+\frac{1}{2}}(x)$ and $V_{k+\frac{1}{2}}(x)$ are given by

$$x_{k,h} = \cos\left(\frac{(2h-1)\pi}{2k+1}\right), \quad (h=1,2,\ldots,k), \tag{25}$$

and the zeros of the pseudo-Chebyshev functions $U_{k+\frac{1}{2}}(x)$ and $W_{k+\frac{1}{2}}(x)$ are given by

$$x_{k,h} = \cos\left(\frac{2h\pi}{2k+1}\right), \quad (h=1,2,\ldots,k). \tag{26}$$

Furthermore, the $W_{k+\frac{1}{2}}(x)$ functions always vanish at the end of the interval $[-1,1]$.

5.4. Hypergeometric Representations

Theorem 5. *The hypergeometric representations of the pseudo-Chebyshev functions are given by:*

$$T_{k+\frac{1}{2}}(x) = \sqrt{\frac{1+x}{2}}\, {}_2F_1\left(-k, k+1, \frac{1}{2}\,\Big|\,\frac{1-x}{2}\right),$$

$$U_{k-\frac{1}{2}}(x) = \frac{2k+1}{1+x}\sqrt{\frac{1}{2(1-x)}}\, {}_2F_1\left(-k, k+1, \frac{3}{2}\,\Big|\,\frac{1-x}{2}\right),$$

$$V_{k+\frac{1}{2}}(x) = \frac{2k+1}{1-x}\sqrt{\frac{1}{2(1+x)}}\, {}_2F_1\left(-k, k+1, \frac{1}{2}\,\Big|\,\frac{1-x}{2}\right),\ and$$

$$W_{k+\frac{1}{2}}(x) = (2k+1)\sqrt{\frac{1-x}{2}}\, {}_2F_1\left(-k, k+1, \frac{3}{2}\,\Big|\,\frac{1-x}{2}\right).$$
(27)

Proof. Equations (27) follow from the hypergeometric representations of the third and fourth kind Chebyshev polynomials (see [16]):

$$V_k(x) = {}_2F_1\left(-k, k+1, \frac{1}{2}\,\Big|\,\frac{1-x}{2}\right),$$

$$W_k(x) = (2k+1)\, {}_2F_1\left(-k, k+1, \frac{3}{2}\,\Big|\,\frac{1-x}{2}\right),$$
(28)

by using Definitions (2). □

5.5. Rodrigues-Type Formulas

Theorem 6. *The Rodrigues-type formulas for the pseudo-Chebyshev functions are given by:*

$$T_{k+\frac{1}{2}}(x) = \frac{(-1)^k}{(2k-1)!!}\sqrt{\frac{1-x}{2}}\,\frac{d^k}{dx^k}\left[(1-x)^{k-1/2}(1+x)^{k+1/2}\right],$$

$$U_{k-\frac{1}{2}}(x) = \frac{(-1)^k}{(2k-1)!!}\frac{1}{1-x}\sqrt{\frac{1}{2(1+x)}}\,\frac{d^k}{dx^k}\left[(1-x)^{k+1/2}(1+x)^{k-1/2}\right],$$

$$V_{k+\frac{1}{2}}(x) = \frac{(-1)^k}{(2k-1)!!}\frac{1}{1+x}\sqrt{\frac{1}{2(1-x)}}\,\frac{d^k}{dx^k}\left[(1-x)^{k-1/2}(1+x)^{k+1/2}\right],\ and$$

$$W_{k+\frac{1}{2}}(x) = \frac{(-1)^k}{(2k-1)!!}\sqrt{\frac{1+x}{2}}\,\frac{d^k}{dx^k}\left[(1-x)^{k+1/2}(1+x)^{k-1/2}\right].$$
(29)

Proof. Equations (29) follow from the Rodrigues-type formulas of the third and fourth kind Chebyshev polynomials (see [16]):

$$V_k(x) = \frac{(-1)^k}{(2k-1)!!}\sqrt{\frac{1-x}{1+x}}\,\frac{d^k}{dx^k}\left[(1-x)^{k-1/2}(1+x)^{k+1/2}\right],$$

$$W_k(x) = \frac{(-1)^k}{(2k-1)!!}\sqrt{\frac{1+x}{1-x}}\,\frac{d^k}{dx^k}\left[(1-x)^{k+1/2}(1+x)^{k-1/2}\right].$$
(30)

by using Definitions (2). □

6. Links with First and Second Kind Chebyshev Polynomials

Theorem 7. *The pseudo-Chebyshev functions are connected with the first and second kind Chebyshev polynomials by means of the equations:*

$$T_{k+\frac{1}{2}}(x) = T_{2k+1}\left(\sqrt{\tfrac{1+x}{2}}\right) = T_{2k+1}\left(T_{1/2}(x)\right),$$

$$U_{k-\frac{1}{2}}(x) = \tfrac{1}{1+x}\sqrt{\tfrac{1}{2(1-x)}}\, U_{2k}\left(\sqrt{\tfrac{1+x}{2}}\right),$$

$$V_{k+\frac{1}{2}}(x) = \tfrac{1}{1-x^2}\, T_{2k+1}\left(\sqrt{\tfrac{1+x}{2}}\right) = \tfrac{1}{1-x^2}\, T_{k+\frac{1}{2}}(x),\ \text{and}$$

$$W_{k+\frac{1}{2}}(x) = \sqrt{\tfrac{1-x}{2}}\, U_{2k}\left(\sqrt{\tfrac{1+x}{2}}\right) = (1-x^2)\, U_{k-\frac{1}{2}}(x).$$

(31)

Proof. The results follow from the equations:

$$V_k(x) = \sqrt{\tfrac{2}{1+x}}\, T_{2k+1}\left(\sqrt{\tfrac{1+x}{2}}\right)\ \text{and}$$

$$W_k(x) = U_{2k}\left(\sqrt{\tfrac{1+x}{2}}\right),$$

(32)

(see [16]), by using Definitions (2). □

Remark 1. *Note that the first equation in (31) is a generalization of the known nesting property, satisfied by the first kind Chebyshev polynomials:*

$$T_m\left(T_n(x)\right) = T_{mn}(x).$$

(33)

This property actually holds in general, independently of the indexes, as a consequence of the basic definition $T_k(x) = \cos(k \arccos(x))$. Note that this composition identity still holds for the first kind Chebyshev polynomials in several variables [4].

7. Conclusions

We have derived the main properties satisfied by the first, second, third, and fourth kind pseudo-Chebyshev polynomials of half-integer degree, which are actually irrational functions. The relevant properties and graphs of these new functions have been derived from their link with the third and fourth kind classical Chebyshev polynomials.

Author Contributions: The authors claim to have contributed equally and significantly in this paper. All authors read and approved the final manuscript.

Funding: The authors declare that they have not received funds from any institution.

Conflicts of Interest: The authors declare no conflict of interest.

References

1. Rivlin, T.J. *The Chebyshev Polynomials*; Wiley: Hoboken, NJ, USA, 1974.
2. Ricci, P.E. Alcune osservazioni sulle potenze delle matrici del secondo ordine e sui polinomi di Tchebycheff di seconda specie. *Atti della Accademia delle Scienze di Torino Classe di Scienze Fisiche Matematiche e Naturali* **1975**, *109*, 405–410.
3. Ricci, P.E. Sulle potenze di una matrice. *Rend. Mat.* **1976**, *9*, 179–194.

4. Ricci, P.E. Una proprietà iterativa dei polinomi di Chebyshev di prima specie in più variabili. *Rend. Mat. Appl.* **1986**, *6*, 555–563.
5. Cesarano, C. Identities and generating functions on Chebyshev polynomials. *Georgian Math. J.* **2012**, *19*, 427–440. [CrossRef]
6. Cesarano, C. Integral representations and new generating functions of Chebyshev polynomials. *Hacet. J. Math. Stat.* **2015**, *44*, 535–546. [CrossRef]
7. Cesarano, C.; Cennamo, G.M.; Placidi, L. Operational methods for Hermite polynomials with applications. *WSEAS Trans. Math.* **2014**, *13*, 925–931.
8. Cesarano, C. Generalized Chebyshev polynomials. *Hacet. J. Math. Stat.* **2014**, *43*, 731–740.
9. Cesarano, C.; Fornaro, C. Operational identities on generalized two-variable Chebyshev polynomials. *Int. J. Pure Appl. Math.* **2015**, *100*, 59–74. [CrossRef]
10. Cesarano, C.; Fornaro, C. A note on two-variable Chebyshev polynomials. *Georgian Math. J.* **2017**, *243*, 339–349. [CrossRef]
11. Mason, J.C.; Handscomb, D.C. *Chebyshev Polynomials*; Chapman and Hall: New York, NY, USA; CRC: Boca Raton, FL, USA, 2003.
12. Boyd, J.P. *Chebyshev and Fourier Spectral Methods*, 2nd ed.; Dover: Mineola, NY, USA, 2001.
13. Srivastava, H.M.; Ricci, P.E.; Natalini, P. A Family of Complex Appell Polynomial Sets. *Rev. Real Acad. Sci. Cienc. Exactas Fis. Nat. Ser. A Math.* **2018**. [CrossRef]
14. Ricci, P.E. Complex spirals and pseudo-Chebyshev polynomials of fractional degree. *Symmetry* **2018**, *10*, 671. [CrossRef]
15. Gielis, J. *The Geometrical Beauty of Plants*; Atlantis Press: Paris, France; Springer: Berlin/Heidelberg, Germany, 2017.
16. Aghigh, K.; Masjed-Jamei, M.; Dehghan, M. A survey on third and fourth kind of Chebyshev polynomials and their applications. *Appl. Math. Comput.* **2008**, *199*, 2–12. [CrossRef]
17. Doha, E.H.; Abd-Elhameed, W.M.; Alsuyuti, M.M. On using third and fourth kinds Chebyshev polynomials for solving the integrated forms of high odd-order linear boundary value problems. *J. Egypt. Math. Soc.* **2015**, *24*, 397–405. [CrossRef]
18. Kim, T.; Kim, D.S.; Dolgy, D.V.; Kwon, J. Sums of finite products of Chebyshev polynomials of the third and fourth kinds. *Adv. Differ. Equ.* **2018**, *2018*, 283. [CrossRef]

© 2019 by the authors. Licensee MDPI, Basel, Switzerland. This article is an open access article distributed under the terms and conditions of the Creative Commons Attribution (CC BY) license (http://creativecommons.org/licenses/by/4.0/).

Article

Numerical Analysis of Boundary Layer Flow Adjacent to a Thin Needle in Nanofluid with the Presence of Heat Source and Chemical Reaction

Siti Nur Alwani Salleh [1,*], Norfifah Bachok [1,2], Norihan Md Arifin [1,2] and Fadzilah Md Ali [1,2]

[1] Institute for Mathematical Research, Universiti Putra Malaysia, UPM Serdang 43400, Selangor, Malaysia; norfifah@upm.edu.my (N.B.); norihana@upm.edu.my (N.M.A.); fadzilahma@upm.edu.my (F.M.A.)
[2] Department of Mathematics, Faculty of Science, Universiti Putra Malaysia, UPM Serdang 43400, Selangor, Malaysia
* Correspondence: alwani24salleh@gmail.com; Tel.: +603-8946-6849

Received: 26 January 2019; Accepted: 22 February 2019; Published: 15 April 2019

Abstract: The steady boundary layer flow of a nanofluid past a thin needle under the influences of heat generation and chemical reaction is analyzed in the present work. The mathematical model has been formulated by using Buongiornos's nanofluid model which incorporates the effect of the Brownian motion and thermophoretic diffusion. The governing coupled partial differential equations are transformed into a set of nonlinear ordinary differential equations by using appropriate similarity transformations. These equations are then computed numerically through MATLAB software using the implemented package called bvp4c. The influences of various parameters such as Brownian motion, thermophoresis, velocity ratio, needle thickness, heat generation and chemical reaction parameters on the flow, heat and mass characteristics are investigated. The physical characteristics which include the skin friction, heat and mass transfers, velocity, temperature and concentration are further elaborated with the variation of governing parameters and presented through graphs. It is observed that the multiple (dual) solutions are likely to exist when the needle moves against the direction of the fluid flow. It is also noticed that the reduction in needle thickness contributes to the enlargement of the region of the dual solutions. The determination of the stable solution has been done using a stability analysis. The results indicate that the upper branch solutions are linearly stable, while the lower branch solutions are linearly unstable. The study also revealed that the rate of heat transfer is a decreasing function of heat generation parameter, while the rate of mass transfer is an increasing function of heat generation and chemical reaction parameters.

Keywords: numerical analysis; heat generation; chemical reaction; thin needle; nanofluid

1. Introduction

In recent decades, the performance of heat transfer of conventional fluids like ethylene glycol, lubricants, oil, kerosene and water, etc., has become less favorable in certain applications. Hence, new kinds of fluid are needed to reach the thermal efficiency for heat exchangers in the future. Choi [1] came out with a tactful idea to resolve the problem by adding dilute suspension of nanoparticles into conventional fluids and this mixture is known as 'nanofluid'. Normally, nanoparticles consist of metals, carbides, oxides, nitrides or non-metals and have dimensions from 1 to 100 nm. Due to the tiny size of nanoelements, nanofluids possess strong suspension stability and able to move without clogging the flow system. Since nanoparticles have higher thermal conductivity than the base fluid, nanofluids are regarded as better coolants particularly in nuclear reactors, domestic refrigerators, transportations, cancer therapy, microelectronic devices, lubricants and also thin film solar energy collectors. A comprehensive literature on the nanofluid applications can be found in the works by Wong and Leon [2], Saidur et al. [3], Huminic and Huminic [4] and Colangelo et al. [5].

In view of above relevant applications, many researchers started to employ nanofluid as an alternative way to enhance the heat transfer efficacy. For instance, Buongiorno [6] established the non-homogeneous equilibrium model which is comprised of seven slip mechanisms; thermophoresis, Brownian diffusion, diffusiophoresis, inertia, gravity Magnus effect and fluid drainage. In this model, the Brownian movement and thermophoretic diffusion of nanoparticles are two notable effects that enhances the thermal conductivity of ordinary liquids. A year after the published work of Buongiorno [6], Tiwari and Das [7] proposed a homogeneous model which taking into account the effect of nanopartilces volume fractions. It is reported that the boundary layer flow of nanofluids over a stretching surface has been studied by Khan and Pop [8]. Makinde and Aziz [9] investigated the boundary layer flow of a nanofluid towards a stretching sheet with convective boundary conditions. Some relevant works on the homogeneous and non-homogeneous models can be seen in the references [10–17].

The study of chemical reactions has amazingly increased due to their wide range useful industrial and technological applications in polymer processing and electrochemistry. Such applications include chemical processing equipment, glass manufacturing, creation and distribution of fog, food processing, energy transfer in a wet cooling tower and evaporation at the surface of the water body [18–20]. The consideration of mixed convection flow past a vertical surface implanted in a porous medium carries species that are relatively soluble in the fluid. In fact, chemical reactions occur due to the presence of a foreign mass in a fluid. In many chemical reactions, the reaction rate relies on the concentration of the species itself. A chemical reactions between the conventional liquid and nanoparticles can be classified as a homogeneous reaction or heterogeneous reaction. Homogeneous reaction is a chemical reaction that occurs consistency in a single phase (gaseous, liquid, or solid). In addition, a heterogeneous reaction is a reaction that involves two or more phases (solid and gas, solid and liquid, two immiscible liquids) and takes place within the boundary of a phase. It is worth mentioning that a chemical reaction is said to be a first order reaction if the reaction rate is directly proportional to the concentration [21,22]. Some applications for the diffusion of species in the boundary layer flow include fibrous insulation, pollution studies and oxidation and synthesis materials. Furthermore, Mabood et al. [23] investigated the influence of chemical reaction on magnetohydrodynamics (MHD) stagnation point flow of nanofluid in porous medium by considering the additional effects of viscous dissipation and thermal radiation. Eid [24] analyzed the chemical reaction effect on MHD nanofluid flow past a stretching sheet with heat generation. It is noticed from his study that the presence of heat source and chemical reaction decrease the heat transfer rate and increase the mass transfer rate. The influence of chemical reaction and heat generation on mixed convection flow of a Casson nanofluid towards a permeable stretching sheet has been studied by Ibrahim et al. [25]. Inspired by the previous works, many authors have considered the chemical reaction effects on different flow concepts as can be seen in the references [26–29]. Very recently, Hayat et al. [30] discussed the mixed convection flow of Williamson nanofluid subject to chemical reaction.

Moreover, the boundary layer flow over a thin needle is of considerable importance in the biomedical and engineering purposes. For instance, it is commonly used in hot wire anemometer or protected thermocouple for calculating the wind velocity, transportations, circulatory problems and wire coating. The topic of thin needle seems very famous due to the movement of the needle that distracts the free-stream flow. This criterion is a primary point of the flow and heat transfer process to calculate the velocity and temperature distributions in experimental studies. Thin (or slender) needle is categorized as a rebellious body whose thicknesses are comparable to that of boundary layer or smaller. The boundary layer development adjacent to a thin needle in viscous fluid is first considered by Lee [31]. Narain and Uberoi [32] analyzed free and mixed convection flow along a thin needle. In extension to which, many works regarding slender needle in a viscous fluid are found in the existing literature [33–36]. Furthermore, the literature shows that researchers have also devoted their attention to the study of boundary layer flow near a slender needle in nanofluid. These situations are caused by the usage of nanofluid that enhance the heat transfer rate. In 2011, the study of the forced convection flow with variable surface temperature over a slender needle has been done by Grosan and Pop [37]. A collection of the boundary layer flow over a thin needle with various physical effects in nanofluid can be found in the work by Trimbitas et al. [38], Hayat et al. [39], Soid et al. [40], Krishna et al. [41] and Ahmad et al. [42]. Very recently, Salleh et al. [43] performed the numerical analysis of magnetohydrodynamics flow over a moving vertical slender needle in nanofluid using Buongiorno's model with the revised boundary conditions.

Therefore, the novelty of the present work is to analyze the problem of the steady laminar nanofluid flow adjacent to a slender needle by considering the additional effects of chemical reaction and also heat source. Buongiorno's model is chosen in the simulation of the nanofluid. The system of nonlinear ordinary differential equations is computed numerically using bvp4c package in MATLAB software. The graphical results are presented and discussed for the varying effect of emerging parameters.

2. Governing Formula and Modeling

A steady nanofluid flow past a horizontal thin needle is examined. The geometry of the problem is illustrated in Figure 1 with u and v denoting x and r components of velocity, respectively, and $r = R(x) = (vcx/U)^{1/2}$ represents the needle radius. The needle is considered to move with uniform velocity U_w in the same or reverse direction of the external flow of constant velocity U_∞ with the composite velocity $U = U_w + U_\infty$. It is assumed that T_w and C_w are the constant wall temperature and nanoparticle concentration and as $r \to \infty$, the ambient temperature and nanoparticle fraction are T_∞ and C_∞ such that $T_w > T_\infty$ and $C_w > C_\infty$. In view of small needle size, the pressure gradient is ignored, however, the transverse curvature effect is required.

By using Buongiorno's nanofluid model, the relevant governing boundary layer systems for the flow are [6,42]

$$\frac{\partial}{\partial x}(ru) + \frac{\partial}{\partial r}(rv) = 0, \tag{1}$$

$$u\frac{\partial u}{\partial x} + v\frac{\partial u}{\partial r} = \frac{v}{r}\frac{\partial}{\partial r}\left(r\frac{\partial u}{\partial r}\right), \tag{2}$$

$$u\frac{\partial T}{\partial x} + v\frac{\partial T}{\partial r} = \frac{\alpha}{r}\frac{\partial}{\partial r}\left(r\frac{\partial T}{\partial r}\right) + \kappa\left[D_B\frac{\partial T}{\partial r}\frac{\partial C}{\partial r} + \frac{D_T}{T_\infty}\left(\frac{\partial T}{\partial r}\right)^2\right] + \frac{Q^*}{\rho C_p}(T - T_\infty), \tag{3}$$

$$u\frac{\partial C}{\partial x} + v\frac{\partial C}{\partial r} = \frac{D_B}{r}\frac{\partial}{\partial r}\left(r\frac{\partial C}{\partial r}\right) + \frac{D_T}{T_\infty}\frac{1}{r}\frac{\partial}{\partial r}\left(r\frac{\partial T}{\partial r}\right) - K^*(C - C_\infty). \tag{4}$$

The physical boundary restrictions are

$$u = U_w, \quad v = 0, \quad T = T_w, \quad C = C_w \quad \text{at} \quad r = R(x),$$

$$u \to U_\infty, \quad T \to T_\infty, \quad C \to C_\infty \quad \text{as} \quad r \to \infty, \tag{5}$$

in which v is kinematic viscosity, T is the temperature of nanofluid, C is the concentration of nanoparticles, α is the thermal diffusivity, ρ is the density, C_p is the heat capacity at uniform pressure, $\kappa = (\rho C_p)_s/(\rho C_p)_f$ is the proportion of effectual heat capacity of nanofluid in which subscripts 's' and 'f' refer to solid nanoparticle and base fluid, $Q^* = Q_0/x$ is the dimensionless heat generation, $K^* = K_0/x$ is the dimensionless reaction rate, Q_0 is the heat generation coefficient and K_0 is the chemical reaction coefficient. It is worth mentioning that the dimensionless parameters Q^* and K^* are the function of x and its value varies locally throughout the flow motion. Besides, D_B and D_T are Brownian and thermophoresis diffusion coefficients, respectively.

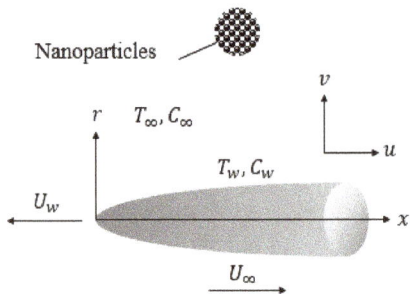

Figure 1. Schematic view of the present study.

The similarity transformation technique has been used for obtaining the ordinary differential equations. Hence, the following non-dimensional parameters are introduced

$$\psi = vxf(\eta), \quad \eta = \frac{Ur^2}{vx}, \quad \theta(\eta) = \frac{T - T_\infty}{T_w - T_\infty}, \quad \phi(\eta) = \frac{C - C_\infty}{C_w - C_\infty}, \tag{6}$$

where the stream functions are given as

$$u = r^{-1}\frac{\partial \psi}{\partial r}, \quad v = -r^{-1}\frac{\partial \psi}{\partial x}. \tag{7}$$

The stream functions (7) satisfies the continuity Equation (1). Using Equations (6) and (7), we obtain the following equations

$$2\eta f''' + 2f'' + ff'' = 0, \tag{8}$$

$$\frac{2}{Pr}(\eta \theta')' + f\theta' + 2\eta \left(Nt\theta'^2 + Nb\theta'\phi'\right) + \frac{1}{2}Q\theta = 0, \tag{9}$$

$$2(\eta \phi')' + 2\frac{Nt}{Nb}(\eta \theta')' + Lef\phi' - \frac{1}{2}LeK\phi = 0. \tag{10}$$

Also, the boundary condition could be rewritten as

$$f(c) = \frac{\varepsilon}{2}c, \quad f'(c) = \frac{\varepsilon}{2}, \quad \theta(c) = 1, \quad \phi(c) = 1,$$

$$f'(\eta) \to \frac{1}{2}(1-\varepsilon), \quad \theta(\eta) \to 0, \quad \phi(\eta) \to 0 \text{ as } \eta \to \infty, \tag{11}$$

where prime denotes the differentiation with regard to similarity variable η. Besides, assume $\eta = c$ to represent size or thickness of the needle.

Here, Pr, Nt, Nb, Q, Le and ε represent the Prandtl number, thermophoresis parameter, Brownian motion parameter, heat generation parameter, Lewis number and velocity ratio parameter. K is the chemical reaction parameter with $K > 0$ represents a destructive reaction, and $K < 0$ represents generative reaction. These non-dimension parameters are defined as follows:

$$Pr = \frac{v}{\alpha}, \quad Nt = \frac{\kappa D_T(T_w - T_\infty)}{vT_\infty}, \quad Nb = \frac{\kappa D_B C_\infty}{v}, \quad Q = \frac{Q_0}{\rho C_p U}, \quad Le = \frac{v}{D_B},$$

$$K = \frac{K_0}{U}, \quad \varepsilon = \frac{U_w}{U}, \tag{12}$$

The skin friction coefficient C_f, local Nusselt number Nu_x and local Sherwood number Sh_x that relate to the shear stress, heat transfer rate and mass transfer rate are defined as

$$C_f = \frac{\mu}{\rho U^2}\left(\frac{\partial u}{\partial r}\right)_{r=c} = 4Re_x^{-1/2}c^{1/2}f''(c), \quad (13)$$

$$Nu_x = \frac{x}{(T_w - T_\infty)}\left(\frac{\partial T}{\partial r}\right)_{r=c} = -2Re_x^{1/2}c^{1/2}\theta'(c), \quad (14)$$

$$Sh_x = \frac{-x}{C_\infty}\left(\frac{\partial C}{\partial r}\right)_{r=c} = -2Re_x^{1/2}c^{1/2}\phi'(c). \quad (15)$$

where $Re_x = Ux/\nu$ is the local Reynold number.

3. Stability Analysis

The idea of the stability analysis came from Weidman et al. [44]. In their study, they noticed that there exists more than one solution called dual solutions. It is important to note that this analysis is introduced to determine which solution provides a good physical meaning to the flow (stable solution). Since we obtained the dual solutions, thus we are encouraged to determine which solutions are stable. To carry out this analysis, Equations (2)–(4) must be in unsteady case. Hence, the new dimensionless time variable is taken as $\tau = 2Ut/x$. Thus, we have

$$\frac{\partial u}{\partial t} + u\frac{\partial u}{\partial x} + v\frac{\partial u}{\partial r} = \frac{\nu}{r}\frac{\partial}{\partial r}\left(r\frac{\partial u}{\partial r}\right), \quad (16)$$

$$\frac{\partial T}{\partial t} + u\frac{\partial T}{\partial x} + v\frac{\partial T}{\partial r} = \frac{\alpha}{r}\frac{\partial}{\partial r}\left(r\frac{\partial T}{\partial r}\right) + \kappa\left[D_B\frac{\partial T}{\partial r}\frac{\partial C}{\partial r} + \frac{D_T}{T_\infty}\left(\frac{\partial T}{\partial r}\right)^2\right] + \frac{Q^*}{\rho C_p}(T - T_\infty), \quad (17)$$

$$\frac{\partial C}{\partial t} + u\frac{\partial C}{\partial x} + v\frac{\partial C}{\partial r} = \frac{D_B}{r}\frac{\partial}{\partial r}\left(r\frac{\partial C}{\partial r}\right) + \frac{D_T}{T_\infty}\frac{1}{r}\frac{\partial}{\partial r}\left(r\frac{\partial T}{\partial r}\right) - K^*(C - C_\infty), \quad (18)$$

and the new similarity transformations take the following form

$$\psi = \nu x f(\eta, \tau), \quad \eta = \frac{Ur^2}{\nu x}, \quad \theta(\eta, \tau) = \frac{T - T_\infty}{T_w - T_\infty}, \quad \phi(\eta, \tau) = \frac{C - C_\infty}{C_w - C_\infty}, \quad \tau = \frac{2Ut}{x}. \quad (19)$$

Please note that the use of τ is related to an initial value problem that is consistent with the solution that will be attained in practice (physically realizable). Afterwards, incorporating Equation (19) into Equations (16)–(18), we obtains

$$2\frac{\partial}{\partial\eta}\left(\eta\frac{\partial^2 f}{\partial\eta^2}\right) + f\frac{\partial^2 f}{\partial\eta^2} - \frac{\partial^2 f}{\partial\eta\partial\tau} + \tau\frac{\partial f}{\partial\eta}\frac{\partial^2 f}{\partial\eta\partial\tau} - \tau\frac{\partial^2 f}{\partial\eta^2}\frac{\partial f}{\partial\tau} = 0, \quad (20)$$

$$\frac{2}{Pr}\frac{\partial}{\partial\eta}\left(\eta\frac{\partial\theta}{\partial\eta}\right) + f\frac{\partial\theta}{\partial\eta} + 2\eta\left[Nt\left(\frac{\partial\theta}{\partial\eta}\right)^2 + Nb\frac{\partial\theta}{\partial\eta}\frac{\partial\phi}{\partial\eta}\right] + \frac{1}{2}Q\theta - \frac{\partial\theta}{\partial\tau} + \tau\frac{\partial f}{\partial\eta}\frac{\partial\theta}{\partial\tau} - \tau\frac{\partial\theta}{\partial\eta}\frac{\partial f}{\partial\tau} = 0, \quad (21)$$

$$2\frac{\partial}{\partial\eta}\left(\eta\frac{\partial\phi}{\partial\eta}\right) + 2\frac{Nt}{Nb}\frac{\partial}{\partial\eta}\left(\eta\frac{\partial\theta}{\partial\eta}\right) + Le\left(f\frac{\partial\phi}{\partial\eta} - \frac{\partial\phi}{\partial\tau}\right) - \frac{1}{2}LeK\phi + Le\tau\frac{\partial f}{\partial\eta}\frac{\partial\phi}{\partial\tau} - Le\tau\frac{\partial\phi}{\partial\eta}\frac{\partial f}{\partial\tau} = 0, \quad (22)$$

together with the boundary conditions

$$f(c,\tau) = \frac{\varepsilon}{2}c + \tau\frac{\partial f}{\partial\tau}(c,\tau), \quad \frac{\partial f}{\partial\eta}(c,\tau) = \frac{\varepsilon}{2}, \quad \theta(c,\tau) = 1, \quad \phi(c,\tau) = 1,$$

$$\frac{\partial f}{\partial\eta}(\eta,\tau) \to \frac{1}{2}(1-\varepsilon), \quad \theta(\eta,\tau) \to 0, \quad \phi(\eta,\tau) \to 0 \text{ as } \eta \to \infty. \quad (23)$$

Subsequently, we assume [44,45]

$$f(\eta, \tau) = f_0(\eta) + e^{-\gamma\tau}F(\eta, \tau),$$
$$\theta(\eta, \tau) = \theta_0(\eta) + e^{-\gamma\tau}G(\eta, \tau), \qquad (24)$$
$$\phi(\eta, \tau) = \phi_0(\eta) + e^{-\gamma\tau}H(\eta, \tau)$$

in order to specify the stability of solutions $f = f_0(\eta)$, $\theta = \theta_0(\eta)$ and $\phi = \phi_0(\eta)$ which meets the boundary value problem (20)–(23). Also, functions $F(\eta, \tau)$, $G(\eta, \tau)$ and $H(\eta, \tau)$ represent small relative to $f_0(\eta)$, $\theta_0(\eta)$ and $\phi_0(\eta)$, respectively, and γ denotes an unknown eigenvalue parameter.

Then, introducing Equation (24) into Equations (20)–(23) yields the linear eigenvalue equations below:

$$2\left(\eta F_0''\right)' + f_0 F_0'' + f_0'' F_0 + \gamma F_0' = 0, \qquad (25)$$

$$\frac{2}{Pr}\left(\eta G_0'\right)' + f_0 G_0' + F_0 \theta_0' + 2\eta Nb\left(\theta_0' H_0' + \phi_0' G_0'\right) + 4\eta Nt\theta_0' G_0' + \frac{1}{2}QG_0 + \gamma G_0 = 0, \qquad (26)$$

$$2\left(\eta H_0'\right)' + 2\frac{Nt}{Nb}\left(\eta G_0'\right)' + Le\left(f_0 H_0' + F_0 \phi_0'\right) - \frac{1}{2}LeKH_0 + Le\gamma H_0 = 0. \qquad (27)$$

The corresponds boundary conditions for these equations are given by

$$F_0(c) = 0, \quad F_0'(c) = 0, \quad G_0(c) = 0, \quad H_0(c) = 0,$$
$$F_0'(\eta) \to 0, \quad G_0(\eta) \to 0, \quad H_0(\eta) \to 0 \quad \text{as} \quad \eta \to \infty. \qquad (28)$$

Next, to identify an early growth or decomposition of the solution (24), we have to set $\tau = 0$. Hence, functions F, G and H can be expressed as $F_0(\eta)$, $G_0(\eta)$ and $H_0(\eta)$, respectively (see Weidman et al. [44] for the detail).

In the present work, we computed the numerical results for Equations (25)–(27) associated with conditions (28) by using the new boundary condition which is $F_0''(c) = 1$. This condition is obtained by relaxing the condition $F_0'(\eta) \to 0$ as $\eta \to \infty$ (see [46] for details). It is to be noted that the flow will be stable if γ is positive, while the flow will be unstable if γ is negative.

4. Graphical Results and Discussion

In this section, the graphical outputs of our problem are interpreted for various effects of the involved parameters. All the computations have been carried out for a wide range of values of the governing parameters; $c(0.1 \leq c \leq 0.2)$, $\varepsilon(-4.3 \leq \varepsilon \leq 0.8)$, $K(-0.1 \leq K \leq 0.2)$, $Q(0 \leq Q \leq 0.4)$, $Nb(0.1 \leq Nb \leq 0.5)$, $Nt(0.1 \leq Nt \leq 0.5)$ and for a fixed values of $Pr = 2$ and $Le = 1$. Equations (8)–(10) along with the conditions (11) are computed numerically via bvp4c function that implemented in MATLAB software. Besides the shooting method, there is a new effective method for solving the boundary value problem for ordinary differential equations that is bvp4c package. Mathematically, this package uses the finite difference methods, in which the output is attained using an initial guess provided at the starting mesh point and resize the step to obtain the particular certainty. Nevertheless, to use this package, the boundary value problem must reduce to first order system of ordinary differential equations. To validate the accuracy of the present results, we have initially compared our results to those of Ahmad et al. [42] and Salleh et al. [43]. In this respect, Table 1 shows a comparison value of shear stress $f''(c)$ for $\varepsilon = Le = Q = K = 0$ for some of the thickness of the needle c when $Pr = 1$. An excellent agreement is observed in these studies.

The effect of needle thickness c on the velocity, temperature and concentration profiles are graphically presented in Figure 2a–c. It is noticed from the plots that the velocity, temperature and concentration profiles for upper branch solution increase with the increasing value of needle thickness. Similar observation is found for momentum, thermal and concentration boundary layer thicknesses for the upper branch as the c increase. Mathematically, the shape of graphs obtained in these profiles has asymptote behaviors and it fulfills the requirement of boundary condition (11). One can see that an increment in the needle thickness decreases the numerical values of surface shear stress $f''(c)$, heat flux $-\theta'(c)$ and also mass flux $-\phi'(c)$. These phenomena are clearly shown in Figure 3a–c. This situation occurs due to an increase in the momentum, thermal and concentration boundary layer thicknesses on the surface, and consequently decline the shear stress and slow down the heat and mass transfers from the surface to the flow. Physically, the slender surface of the needle makes heat and mass to diffuse through it quickly compared to thick surface. In addition, the critical values of ε, by which the upper and lower branch solutions connected, are noticed to decrease as the needle thickness reduces. In other words, we can say that the needle thickness has a significant effect on the existence of the dual solutions.

Table 1. Comparison values of shear stress $f''(c)$ when $\varepsilon = Le = Q = K = 0$ for some of the thickness of the needle c when $Pr = 1$.

c	Ahmad et al. [42]	Salleh et al. [43]	Current Study
0.01	8.4924360	8.4924452	8.4924453
0.1	1.2888171	1.2888299	1.2888300
0.15	-	-	0.9383388
0.2	-	-	0.7515725

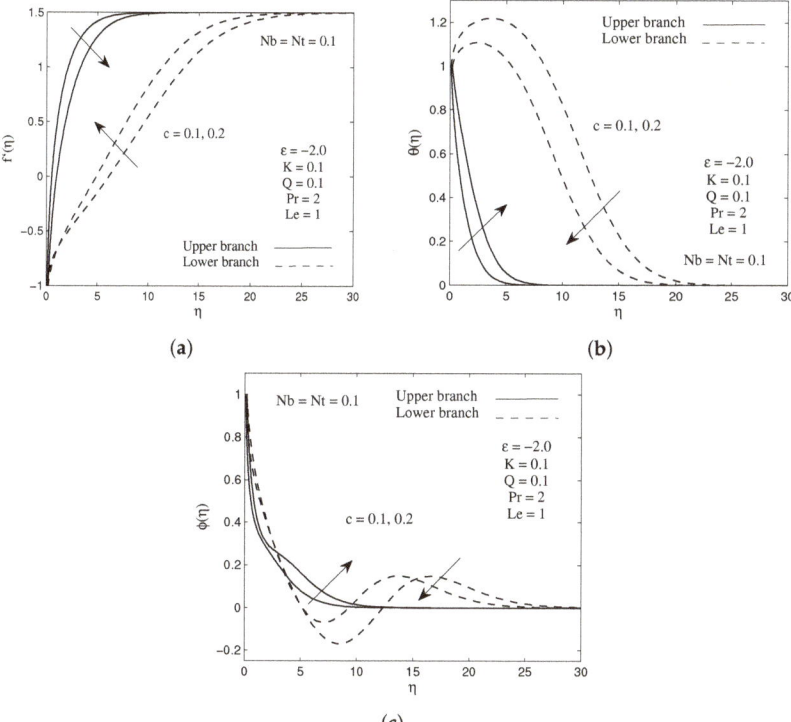

Figure 2. Sample of (**a**) velocity, (**b**) temperature and (**c**) concentration profiles for several values of needle thickness c.

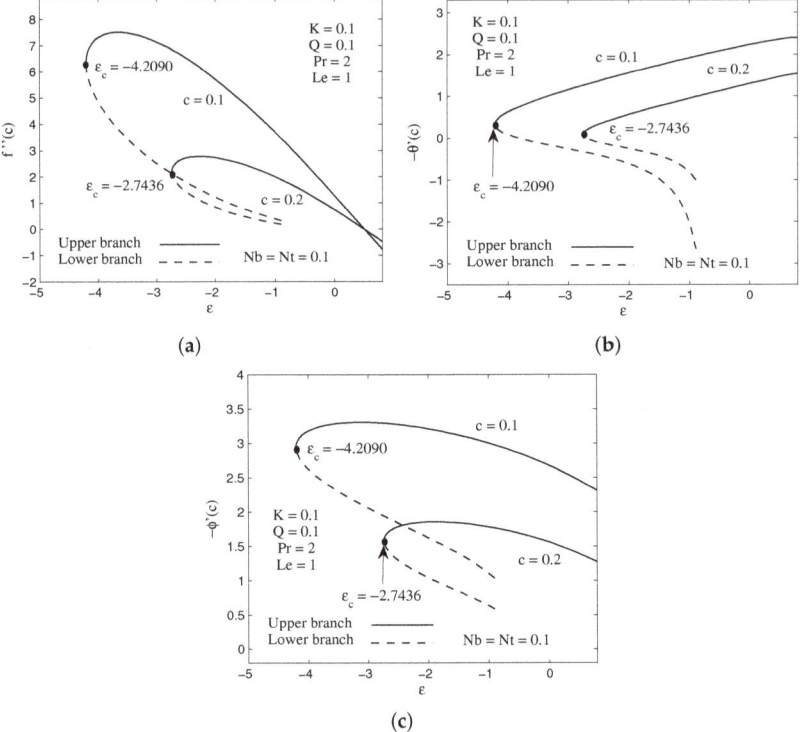

Figure 3. Variation of surface (**a**) shear stress, (**b**) local heat flux and (**c**) local mass flux with velocity ratio parameter ε for several values of needle thickness c.

The distributions of the temperature and concentration for several values of heat generation parameter Q are illustrated in Figure 4a–b when $K = 0.1$. It is worth noticing from Figure 4a that the fluid temperature enhances significantly within the thermal boundary layer for higher values of heat generation parameter. However, the opposite effect is observed for the fluid concentration as the heat generation parameter increases. This characteristic can be seen in Figure 4b. Figure 5a–b visualize the effect of the heat generation parameter Q on the local heat and mass fluxes with velocity ratio parameter ε. It is found from the figures that increasing values of Q reduce the heat flux at the surface, while an opposite criterion is observed for the mass flux. Generally, the presence of the heat generator produces a hot fluid layer near the needle surface due to mechanism of heat generation. As a consequence, the rate of heat transfer decreases from the needle surface to the fluid flow. In addition, the decrement of heat transfer is also due to an increment in the thermal boundary layer thickness as Q increases (see Figure 4a). Moreover, the existence of the hot fluid in the system will accelerate the motion of nanoparticles, and as the result increases the rate of mass transfer on the needle surface. It is worth mentioning that the existence of the dual solutions is noted when the needle moves against the free stream direction, $\varepsilon < 0$. In these variations, the presence of heat generation does not affect the flow. This statement can be proved by looking at Equation (8), where it does not contain the parameter Q inside.

The concentration distributions for some values of chemical reaction parameter K are plotted in Figure 6 when $Q = 0.1$. It is discerned from this figure that the fluid concentration is decreasing function of the chemical reaction parameter. It is important to know that the nanoparticle concentration as well as boundary layer thickness diminishes with the destructive chemical reaction, $K > 0$. Noteworthy, with the existence of destructive reaction, the change of the species as a cause of chemical reaction will reduce the nanoparticle concentration in the boundary layer thickness. Other than that, Figure 7 elucidates the influence of chemical reaction parameter on the local mass flux. This figure explains that increasing the chemical reaction parameter K results in an increase in the mass transfer rate on the surface. The reason behind this is that the reduction of the concentration boundary layer thickness causes the mass transfer takes place quickly between the needle surface and the fluid flow. Since the chemical reaction parameter exists in Equation (10), thus we present only the result for the mass flux here.

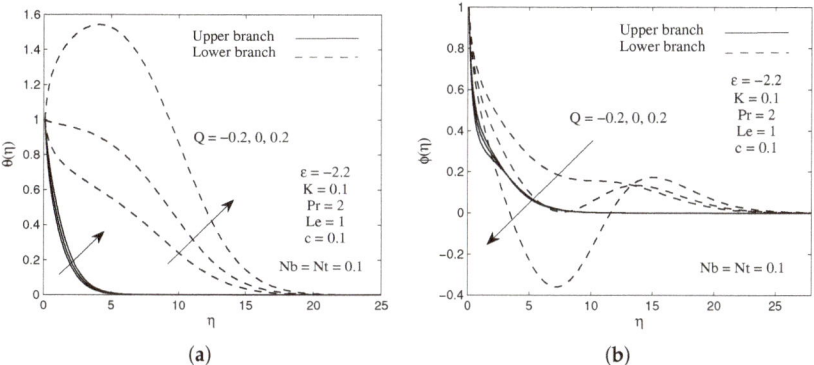

Figure 4. Sample of (**a**) temperature and (**b**) concentration profiles for several values of heat generation parameter Q.

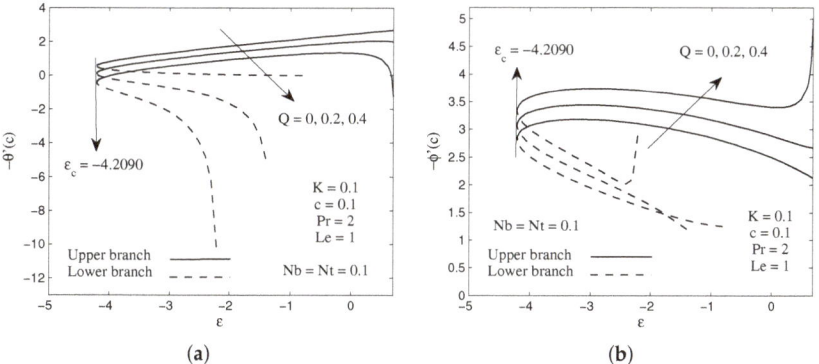

Figure 5. Variation of surface (**a**) local heat flux and (**b**) local mass flux with velocity ratio parameter ε for several values of heat generation parameter Q.

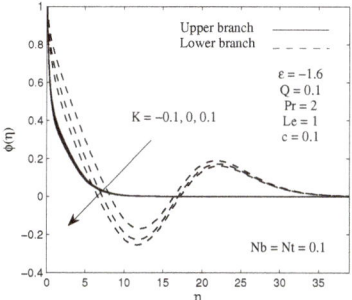

Figure 6. Sample of concentration profiles for several values of chemical reaction parameter K.

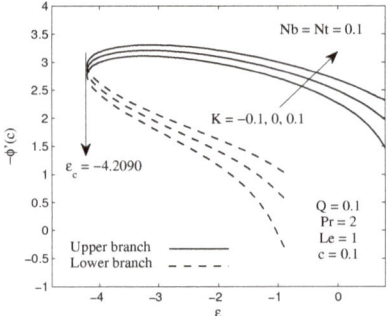

Figure 7. Variation of surface local mass flux for several values of chemical reaction parameter K.

The variations of the local Nusselt number $(Re_x)^{-1/2}Nu_x$ and local Sherwood number $(Re_x)^{-1/2}Sh_x$ with thermophoresis parameter Nt and Brownian motion parameter Nb are presented in Tables 2 and 3. Table 2 indicates that the higher values of Nt and Nb decrease the local Nusselt number or rate of heat transfer occurs on the needle surface. The same features can be seen as the heat generation parameter increases. It is quite clear from Figures 8a and 9a that the higher rate of Brownian motion and thermophoresis enhance the temperature of the fluid as well as the thermal boundary layer thickness. This increment in the boundary layer thickness minimizes the rate of heat transfer from the needle to the flow. Furthermore, Table 3 clarifies that the rate of mass transfer (or local Sherwood number) increases with an increase in thermophoresis and chemical reaction parameters. Noticeably, the higher value of the Brownian motion parameter tends to slow down the rate of mass transfer in the system. This happens due to the continuous collision of base fluid particles and nanoparticles which cause the random movement of those particles in the fluid. In addition, it can be observed in Figure 8b that the presence of higher value of Nb enhance the concentration profiles as well as the concentration boundary layer thickness. This criterion leads to the decrement in the mass transfer rate. In addition, as thermophoretic effect intensifies, nanoparticles with high thermal conductivity penetrate deeper in the fluid, hence, decreases the concentration boundary layer thickness as well as concentration profiles as can be seen in Figure 9b. This behavior leads to an increase in the local Sherwood Number.

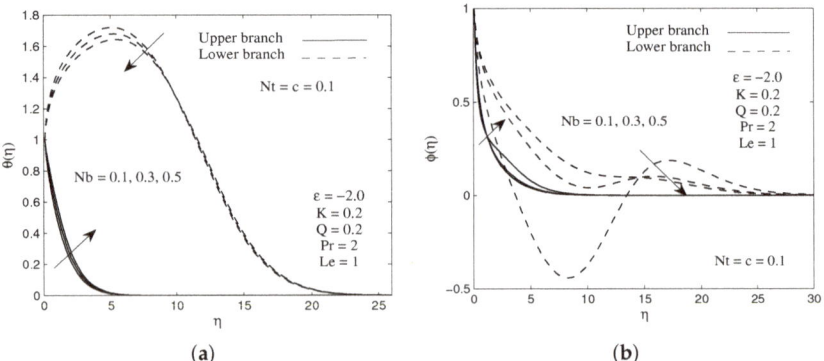

Figure 8. Sample of (a) temperature and (b) concentration profiles for several values of Brownian motion parameter Nb.

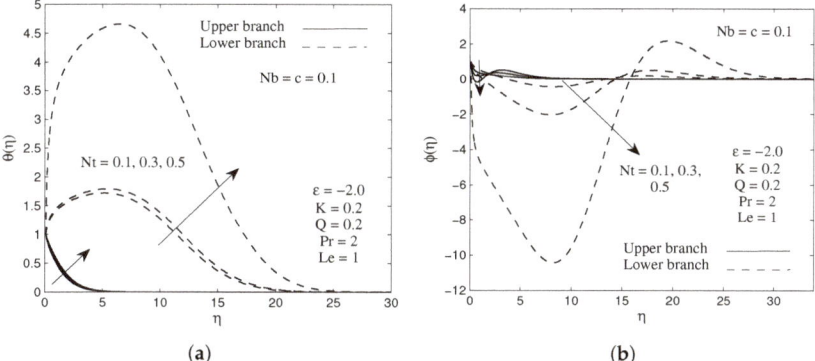

(a) (b)

Figure 9. Sample of (a) temperature and (b) concentration profiles for several values of thermophoresis parameter Nt.

Table 2. Effects of thermphoresis parameter Nt and Brownian motion parameter Nb on the numerical values of local Nusselt number, $(Re_x)^{-1/2}Nu_x$ for $Q = 0.1$ and $Q = 0.2$ when $\varepsilon = -1.0$, $K = 0.2$, $c = 0.1$, $Pr = 2$ and $Le = 1$.

Heat Generation Parameter	Thermphoresis Parameter	$(Re_x)^{-1/2}Nu_x = -2c^{1/2}\theta'(c)$		
		$Nb = 0.1$	$Nb = 0.3$	$Nb = 0.5$
0.1	0.1	1.208880	0.959223	0.749362
	0.3	0.989170	0.777343	0.601318
	0.5	0.805833	0.627349	0.480535
0.2	0.1	1.078738	0.832762	0.628219
	0.3	0.863788	0.656596	0.486510
	0.5	0.685682	0.512462	0.371923

Table 3. Effects of thermphoresis parameter Nt and Brownian motion parameter Nb on the numerical values of local Sherwood number, $(Re_x)^{-1/2}Sh_x$ for $K = 0.1$ and $K = 0.2$ when $\varepsilon = -1.0$, $Q = 0.2$, $c = 0.1$, $Pr = 2$ and $Le = 1$.

Chemical Reaction Parameter	Thermphoresis Parameter	$(Re_x)^{-1/2}Sh_x = -2c^{1/2}\phi'(c)$		
		$Nb = 0.1$	$Nb = 0.3$	$Nb = 0.5$
0.1	0.1	2.005009	1.825444	1.781448
	0.3	3.271205	2.343083	2.135890
	0.5	5.055702	2.990494	2.546082
0.2	0.1	2.085015	1.898138	1.852426
	0.3	3.362939	2.418572	2.207592
	0.5	5.144680	3.063527	2.615317

Since this study has more than one solution, we need to verify which of the solutions obtained are physically relevant (stable solution) by solving Equations (25)–(28). The determination of the stable solution count on the sign of the smallest eigenvalue γ gained through this analysis. Table 4 presents the smallest eigenvalue γ for several values of chemical reaction, heat generation and velocity ratio parameters when $c = 0.1$ and $c = 0.2$. Table 4 indicates that the positive sign of γ for upper branch solution represents an initial decomposition of disturbance, while the negative sign of γ for lower branch solution represents an initial growth of disturbance in the system. Please note that the flow is said to be stable and physically relevant, if there is an initial decay of disturbance in the boundary layer separation. Otherwise, the flow is said to be unstable and not physically relevant.

Table 4. Smallest eigenvalues γ for several values of chemical reaction parameter K, heat generation parameter Q and velocity ratio parameter ε for $c = 0.1$ and $c = 0.2$ when $Nb = Nt = 0.1$, $Pr = 2$ and $Le = 1$.

K = Q	c	ε	Upper Branch	Lower Branch
0.1	0.1	−4.1994	0.0471	−0.0449
		−4.199	0.0481	−0.0458
		−4.19	0.0668	−0.0625
	0.2	−2.7424	0.0150	−0.0147
		−2.742	0.0175	−0.0170
		−2.74	0.0265	−0.0255
0.2	0.1	−4.1246	0.1444	−0.1254
		−4.124	0.1449	−0.1258
		−4.12	0.1484	−0.1284
	0.2	−2.7136	0.0793	−0.0706
		−2.713	0.0801	−0.0713
		−2.71	0.0841	−0.0744

5. Final Remarks

In this work, the numerical model is developed to study the boundary layer flow of two-phase nanofluid on a moving slender needle. The influences of chemical reaction and heat generation on the flow have been taken into consideration. The governing flow equations are solved and validated numerically by applying bvp4c package through MATLAB software. The key findings of this analysis can be summarized as follows:

- The heat generation parameter reduces the local heat flux as well as the rate of heat transfer.
- The presence of a chemical reaction increases the rate of mass transfer on the needle surface.
- The Brownian motion parameter diminishes the rate of heat and mass transfers from the needle surface to the flow.
- An increase in the thermophoresis parameter results in an increase in the mass transfer rate, while the reverse effect is noted for the heat transfer rate.
- An increment in the needle thickness leads to decrease the magnitudes of the surface shear stress, local heat flux and local mass flux.
- The dual solutions are likely to exist when the needle surface moves against the free-stream direction, $\varepsilon < 0$.
- The upper branch solution exhibits stable flow (or solution) and lower branch solution exhibits unstable flow.

Author Contributions: S.N.A.S. and N.B. designed the research; S.N.A.S. formulated the mathematical model and computed the numerical results; S.N.A.S. and N.B. analyzed the results; S.N.A.S. wrote the manuscript; S.N.A.S., N.B., N.M.A. and F.M.A. have read and approved this manuscript.

Funding: This research was funded by Fundamental Research Grant Scheme (FRGS/1/2018/STG06/UPM/02/4/5540155) from Ministry of Higher Education Malaysia and Putra Grant GP-IPS/2018/9667900 from Universiti Putra Malaysia.

Acknowledgments: The authors would like to express their gratitude to the anonymous reviewers for their valuable comments and suggestions for a betterment of this paper.

Conflicts of Interest: The authors declare no conflicts of interest.

Abbreviations

c	Needle size
C	Fluid concentration (kg m^{-3})
C_f	Skin friction coefficient
C_∞	Ambient nanoparticle volume fraction
C_w	Surface volume fraction
C_p	Specific heat at constant pressure
D_B	Brownian diffusion coefficient (m^2 s^{-1})
D_T	Thermophoretic diffusion coefficient (m^2 s^{-1})
f	Similarity function for velocity
K	Chemical reaction parameter
K_0	Chemical reaction coefficient
K^*	Dimensionless reaction rate
Le	Lewis number
Nb	Brownian motion parameter
Nt	Thermophoresis parameter
Nu_x	Local Nusselt number
Pr	Prandtl number
Q	Heat generation parameter
Q_0	Heat generation coefficient
Q^*	Dimensionless heat generation
r	Cartesian coordinate
Re_x	Local Reynolds number
Sh_x	Local Sherwood number
T	Fluid temperature (K)
T_w	Wall temperature (K)
T_∞	Ambient temperature (K)
U	Composite velocity (ms^{-1})
U_w	Wall velocity (ms^{-1})
U_∞	Ambient velocity (ms^{-1})
u	Velocity in x direction (ms^{-1})
v	Velocity in r direction (ms^{-1})
x	Cartesian coordinate
α	Thermal diffusivity (m^2 s^{-1})
η	Similarity independent variable
θ	Dimensionless temperature
ε	Velocity ratio parameter
κ	Ratio of effective heat capacity of nanofluid
ρC_p	Volumetric heat capacity (J K^{-1})
ν	Kinematic viscosity (m^2 s^{-1})
μ	Dynamic viscosity (kg m^{-1}s^{-1})
ρ	Fluid density (kg m^{-3})
ϕ	Dimensionless solid volume fraction
w	Condition at the wall
∞	Ambient condition
$'$	Differentiative with respect to η

References

1. Choi, S.U.S. Enhancing thermal conductivity of fluids with nanoparticles. *Am. Soc. Mech. Eng. Fluids Eng. Div.* **1995**, *231*, 99–105.
2. Wong, K.V.; Leon, O.D. Applications of Nanofluids: Current and Future. *Adv. Mech. Eng.* **2010**, *2010*, 519659. [CrossRef]
3. Saidur, R.; Leong, K.Y.; Mohammad, H.A. A review on applications and challenges of nanofluids. *Renew. Sustain. Energy Rev.* **2011**, *15*, 1646–1668. [CrossRef]

4. Huminic, G.; Huminic, A. Application of nanofluids in heat exchangers: A review. *Renew. Sustain. Energy Rev.* **2012**, *16*, 5625–5638. [CrossRef]
5. Colangelo, G.; Favale, E.; Milanese, M.; Risi, A.D.; Laforgia, D. Cooling of electronic devices: Nanofluids contribution. *Appl. Ther. Eng.* **2017**, *127*, 421–435. [CrossRef]
6. Buongiorno, J. Convective transport in nanofluids. *J. Heat Trans.* **2006**, *128*, 240–250. [CrossRef]
7. Tiwari, R.K.; Das, M.K. Heat transfer augmentation in a two-sided lid-driven differentially heated square cavity utilizing nanofluids. *Int. J. Heat Mass Trans.* **2007**, *50*, 2002–2018. [CrossRef]
8. Khan, W.A.; Pop, I. Boundary-layer flow of a nanofluid past a stretching sheet. *Int. J. Heat Mass Trans.* **2010**, *53*, 2477–2483. [CrossRef]
9. Makinde, O.D.; Aziz, A. Boundary layer flow of a nanofluid past a stretching sheet with a convective boundary condition. *Int. J. Ther. Sci.* **2011**, *50*, 1326–1332. [CrossRef]
10. Bachok, N.; Ishak, A.; Pop, I. Unsteady boundary-layer flow and heat transfer of a nanofluid over a permeable stretching/shrinking sheet. *Int. J. Heat Mass Trans.* **2012**, *55*, 2102–2109. [CrossRef]
11. Das, K.; Duari, P.R.; Kundu, P.K. Nanofluid flow over an unsteady stretching surface in presence of thermal radiation. *Alex. Eng. J.* **2014**, *53*, 737–745. [CrossRef]
12. Mabood, F.; Khan, W.A.; Ismail, A.I.M. MHD boundary layer flow and heat transfer of nanofluids over a nonlinear stretching sheet: A numerical study. *J. Magn. Magn. Mater.* **2015**, *374*, 569–576. [CrossRef]
13. Naramgari, S.; Sulochana, C. MHD flow over a permeable stretching/shrinking sheet of a nanofluid with suction/injection. *Alex. Eng. J.* **2016**, *55*, 819–827. [CrossRef]
14. Pandey, A.K.; Kumar, M. Boundary layer flow and heat transfer analysis on Cu-water nanofluid flow over a stretching cylinder with slip. *Alex. Eng. J.* **2017**, *56*, 671–677. [CrossRef]
15. Mustafa, M. MHD nanofluid flow over a rotating disk with partial slip effects: Buongiorno model. *Int. J. Heat Mass Trans.* **2017**, *108*, 1910–1916. [CrossRef]
16. Jyothi, K.; Reddy, P.S.; Reddy, M.S. Influence of magnetic field and thermal radiation on convective flow of SWCNTs-water and MWCNTs-water nanofluid between rotating stretchable disks with convective boundary conditions. *Adv. Powder Technol.* **2018**, *331*, 326–337. [CrossRef]
17. Bakar, N.A.A.; Bachok, N.; Arifin, N.M.; Pop, I. Stability analysis on the flow and heat transfer of nanofluid past a stretching/shrinking cylinder with suction effect. *Results Phys.* **2018**, *9*, 1335–1344. [CrossRef]
18. Griffith, R.M. Velocity, temperature and concentration distributions during fiber spinning. *Ind. Eng. Chem. Fundam.* **1964**, *3*, 245–250. [CrossRef]
19. Chin, D.T. Mass transfer to a continuousmoving sheet electrode. *J. Electrochem. Soc.* **1975**, *122*, 643–646. [CrossRef]
20. Gorla, R.S.R. Unsteady mass transfer in the boundary layer on a continuous moving sheet electrode. *J. Electrochem. Soc.* **1978**, *125*, 865–869. [CrossRef]
21. Damseh, R.A.; Al-Odat, M.Q.; Chamkha, A.J.; Shannak, B.A. Combined effect of heat generation or absorption and first-order chemical reaction on micropolar fluid flows over a uniformly stretched permeable surface. *Int. J. Therm. Sci.* **2009**, *48*, 1658–1663. [CrossRef]
22. Magyari, E.; Chamkha, A.J. Combined effect of heat generation or absorption and first-order chemical reaction on micropolar fluid flows over a uniformly stretched permeable surface: The full analytical solution. *Int. J. Therm. Sci.* **2010**, *49*, 1821–1828. [CrossRef]
23. Mabood, F.; Shateyi, S.; Rashidi, M.M.; Momoniat, E.; Freidoonimehr, N. MHD stagnation point flow heat and mass transfer of nanofluids in porous medium with radiation, viscous dissipation and chemical reaction. *Adv. Powder Technol.* **2016**, *27*, 742–749. [CrossRef]
24. Eid, M.R. Chemical reaction effect on MHD boundary-layer flow of two-phase nanofluid model over an exponentially stretching sheet with a heat generation. *J. Mol. Liq.* **2016**, *220*, 718–725. [CrossRef]
25. Ibrahim, S.M.; Lorenzini, G.; Vijaya Kumar, P.; Raju, C.S.K. Influence of chemical reaction and heat source on dissipative MHD mixed convection flow of a Casson nanofluid over a nonlinear permeable stretching sheet. *Int. J. Heat Mass Trans.* **2017**, *111*, 346355. [CrossRef]
26. Nayak, M.K.; Akbar, N.S.; Tripathi, D.; Khan, Z.H.; Pandey, V.S. MHD 3D free convective flow of nanofluid over an exponentially stretching sheet with chemical reaction. *Adv. Powder Technol.* **2017**, *28*, 2159–2166. [CrossRef]

27. Sithole, H.; Mondal, H.; Goqo, S.; Sibanda, P.; Motsa, S. Numerical simulation of couple stress nanofluid flow in magneto-porous medium with thermal radiation and a chemical reaction. *Appl. Math. Comput.* **2018**, *339*, 820–836. [CrossRef]
28. Khan, M.; Shahid, A.; Malik, M.Y.; Salahuddin, T. Chemical reaction for Carreau-Yasuda nanofluid flow past a nonlinear stretching sheet considering Joule heating. *Results Phys.* **2018**, *8*, 1124–1130. [CrossRef]
29. Zeeshan, A.; Shehzad, N.; Ellahi, R. Analysis of activation energy in Couette-Poiseuille flow of nanofluid in the presence of chemical reaction and convective boundary conditions. *Results Phys.* **2018**, *8*, 502–512. [CrossRef]
30. Hayat, T.; Kiyani, M.Z.; Alsaedi, A.; Ijaz Khan, M.; Ahmad, I. Mixed convective three-dimensional flow of Williamson nanofluid subject to chemical reaction. *Int. J. Heat Mass Trans.* **2018**, *127*, 422–429. [CrossRef]
31. Lee, L.L. Boundary layer over a thin needle. *Phys. Fluids* **1967**, *10*, 1820–1822. [CrossRef]
32. Narain, J.P.; Uberoi, S.M. Combined forced and free-convection heat transfer from vertical thin needles in a uniform stream. *Phys. Fluids* **1973**, *15*, 1879–1882. [CrossRef]
33. Wang, C.Y. Mixed convection on a vertical needle with heated tip. *Phys. Fluids* **1990**, *2*, 622–625. [CrossRef]
34. Ishak, A.; Nazar, R.; Pop, I. Boundary layer flow over a continuously moving thin needle in a parallel free stream. *Chin. Phys. Lett.* **2007**, *24*, 2895–2897. [CrossRef]
35. Ahmad, S.; Arifin, N.M.; Nazar, R.; Pop, I. Mixed convection boundary layer flow along vertical thin needles: Assisting and opposing flows. *Int. Commun. Heat Mass Trans.* **2008**, *35*, 157–162. [CrossRef]
36. Afridi, M.I.; Qasim, M. Entropy generation and heat transfer in boundary layer flow over a thin needle moving in a parallel stream in the presence of nonlinear Rosseland radiation. *Int. J. Therm. Sci.* **2018**, *123*, 117–128. [CrossRef]
37. Grosan, T.; Pop, I. Forced Convection Boundary Layer Flow Past Nonisothermal Thin Needles in Nanofluids. *J. Heat Trans.* **2011**, *133*. [CrossRef]
38. Trimbitas, R.; Grosan, T.; Pop, I. Mixed convection boundary layer flow along vertical thin needles in nanofluids. *Int. J. Numer. Methods Heat Fluid Flow* **2014**, *24*, 579–594. [CrossRef]
39. Hayat, T.; Khan, M.I.; Farooq, M.; Yasmeen, T.; Alsaedi, A. Water-carbon nanofluid flow with variable heat flux by a thin needle. *J. Mol. Liq.* **2016**, *224*, 786–791. [CrossRef]
40. Soid, S.K; Ishak, A.; Pop, I. Boundary layer flow past a continuously moving thin needle in a nanofluid. *Appl. Therm. Eng.* **2017**, *114*, 58–64. [CrossRef]
41. Krishna, P.M.; Sharma, R.P.; Sandeep, N. Boundary layer analysis of persistent moving horizontal needle in Blasius and Sakiadis magnetohydrodynamic radiative nanofluid flows. *Nucl. Eng. Technol.* **2017**, *49*, 1654–1659. [CrossRef]
42. Ahmad, R.; Mustafa, M.; Hina, S. Buongiorno's model for fluid flow around a moving thin needle in a flowing nanofluid: A numerical study. *Chin. J. Phys.* **2017**, *55*, 1264–1274. [CrossRef]
43. Salleh, S.N.A.; Bachok, N.; Arifin, N.M.; Ali, F.M.; Pop, I. Magnetohydrodynamics flow past a moving vertical thin needle in a nanofluid with stability analysis. *Energies* **2018**, *11*, 3297, doi.org/10.3390/en11123297. [CrossRef]
44. Weidman, P.D.; Kubitschek, D.G.; Davis, A.M.J. The effect of transpiration on self-similar boundary layer flow over moving surfaces. *Int. J. Eng. Sci.* **2006**, *44*, 730–737. [CrossRef]
45. Rosca, A.V.; Pop, I.Flow and heat transfer over a vertical permeable stretching/shrinking sheet with a second order slip. *Int. J. Heat Mass Trans.* **2013**, *60*, 355-364. [CrossRef]
46. Harris, S.D.; Ingham, D.B.; Pop, I. Mixed convection boundary-layer flow near the stagnation point on a vertical surface in a porous medium: Brinkman model with slip. *Transp. Porous Media* **2009**, *77*, 267–285. [CrossRef]

© 2019 by the authors. Licensee MDPI, Basel, Switzerland. This article is an open access article distributed under the terms and conditions of the Creative Commons Attribution (CC BY) license (http://creativecommons.org/licenses/by/4.0/).

Article

k-Hypergeometric Series Solutions to One Type of Non-Homogeneous *k*-Hypergeometric Equations

Shengfeng Li [1,*] **and Yi Dong** [2]

[1] Institute of Applied Mathematics, Bengbu University, Bengbu 233030, China
[2] School of Science, Bengbu University, Bengbu 233030, China; dy@bbc.edu.cn
* Correspondence: lsf@bbc.edu.cn; Tel.: +86-552-317-5158

Received: 9 January 2019; Accepted: 14 February 2019; Published: 19 February 2019

Abstract: In this paper, we expound on the hypergeometric series solutions for the second-order non-homogeneous *k*-hypergeometric differential equation with the polynomial term. The general solutions of this equation are obtained in the form of *k*-hypergeometric series based on the Frobenius method. Lastly, we employ the result of the theorem to find the solutions of several non-homogeneous *k*-hypergeometric differential equations.

Keywords: *k*-hypergeometric differential equations; non-homogeneous; *k*-hypergeometric series; special function; general solution; Frobenius method

1. Introduction

It is well known that many phenomena in physical and technical applications are governed by a variety of differential equations. We should notice that these differential equations have appeared in many different research fields, for instance in the theory of automorphic function, in conformal mapping theory, in the theory of representations of Lie algebras, and in the theory of difference equations. Analytical and numerical methods to solve ordinary differential equations are an ancient and interesting research direction in differentiable dynamical systems and their applications. Let us consider a so-called non-homogeneous *k*-hypergeometric differential equation of the form:

$$kz(1-kz)\frac{d^2y}{dz^2} + [c - (k+a+b)kz]\frac{dy}{dz} - aby = f(z) \qquad (1)$$

with the independent variable z, where a, b, c, k are several constants with $a, b, c \in \mathbb{R}, k \in \mathbb{R}^+$, and the function $f(z)$ is holomorphic in an interval $\mathcal{D} \subseteq \mathbb{C}$. In the case of $k = 1$, if the function $f(z)$ vanishes identically, then Equation (1) degrades into a linear homogeneous hypergeometric ordinary differential equation presented by Euler [1] in 1769, which has the following normalized form:

$$z(1-z)\frac{d^2y}{dz^2} + [c - (1+a+b)z]\frac{dy}{dz} - aby = 0; \qquad (2)$$

such an equation has been extensively studied.

The solutions of a differential equation relate to many absorbing special functions in mathematics, physics, and engineering. For instance, the solution could be presented by power series [2,3], continued fraction [4–6], zeta function [7–10], and hypergeometric series [11–16]. Among these special functions, the hypergeometric series, denoted by:

$$_2F_1[a,b;c;z] = \sum_{n=0}^{\infty} \frac{(a)_n (b)_n}{(c)_n n!} z^n,$$

can be applied to the solution of the differential Equation (2). For Equation (2), a hypergeometric series solution $_2F_1$ can be derived by the Frobenius method. The so-called hypergeometric series was researched firstly by Wallis [11] in 1655. Since then, Euler, too, had researched the topic on the hypergeometric series, but the first full systematic study was introduced by Gauss [12]. Some works and complete references concerning both the hypergeometric series and the certain equation (2) can be found in Kummer [13], Riemann [14], Bailey [15,16], Chaundy [17], Srivastava [18], Whittaker [19], Beukers [20], Gasper [21], Olde Daalhuis [22,23], Dwork [24], Chu [25], Yilmazer et al. [26], Morita et al. [27], Abramov et al. [28], Alfedeel et al. [29], and the literature therein. However, in contrast to the extensive studies on Equation (2), other hypergeometric differential equations with $k \in \mathbb{R}^+$ are very limited.

If k is not necessarily equal to one and $f(z)$ is still a zero function in Equation (1), then the associated differential equation is written as follows:

$$kz(1-kz)\frac{d^2y}{dz^2} + [c - (k+a+b)kz]\frac{dy}{dz} - aby = 0. \tag{3}$$

This differential Equation (3), called the homogeneous k-hypergeometric differential equation, has been defined only in recent years. For $k \in \mathbb{R}^+$ and $f(z) = 0$, Equation (3) has a solution in the form of k-hypergeometric series $_2F_{1,k}$, which will be introduced in Section 2. It is clear that the k-hypergeometric series $_2F_{1,k}$ has evolved from the hypergeometric series $_2F_1$. Hence, we mention the works of Díaz et al. [30,31], Krasniqi [32,33], Kokologiannaki [34], Mubeen et al. (see [35–40]), Rehman et al. [41,42], and the references therein for results on k-hypergeometric series and the homogeneous k-hypergeometric differential equation. In 2005, the Pochhanner k-symbol was developed by Díaz et al. [30]. Since then, for example, k-gamma and k-beta functions have been researched, and their relevant properties have been shown [30,31]. By following the works of Díaz et al., in 2010, some fascinating results with respect to k-gamma, -beta, and -zeta functions were proven in [32–34]. In 2012, a so-called k-fractional integral and its application were presented by Mubeen and Habibullah [36]. Furthermore, based on the properties of Pochhammer k-symbols, k-gamma, and k-beta functions, Mubeen et al. [35,37] suggested an integral representation of k-hypergeometric functions and some generalized confluent k-hypergeometric functions. Mubeen [37] did not introduce the second-order linear k-hypergeometric differential equation defined by Equation (3) until 2013. Furthermore, in 2014, Mubeen et al. [38,39] solved the k-hypergeometric differential equation by using the Frobenius method and gave its solution in the form of the so-called k-hypergeometric series $_2F_{1,k}$ introduced by Díaz et al. [30]. In the case of $k \in \mathbb{R}^+$ and $f(z) \neq 0$, the research for this question is very limited.

Motivated by the above results, in this paper, we consider the k-hypergeometric series solutions of Equation (1) when $f(z)$ is a non-vanishing function and $k \in \mathbb{R}^+$. For simplicity, we choose $f(z)$ as a polynomial $\sum_{i=0}^{m} d_i z^i$. That is, we will discuss the general solution of the so-called non-homogeneous k-hypergeometric equation:

$$kz(1-kz)\frac{d^2y}{dz^2} + [c - (k+a+b)kz]\frac{dy}{dz} - aby = \sum_{i=0}^{m} d_i z^i, \tag{4}$$

where $k \in \mathbb{R}^+$ and $d_i, i = 0, 1, 2, \ldots, m$, are some real or complex constants. The corresponding homogeneous k-hypergeometric equation of Equation (4) is denoted by Equation (3).

The aim of this paper to find general solutions of the non-homogeneous k-hypergeometric Equation (4) by means of the k-hypergeometric series. This paper is organized as follows: in Section 2, the basic definitions and facts of the k-hypergeometric series and ordinary differential equation are presented. Our results are then introduced in Section 3. Some examples are given to illustrate the applications of our results in Section 4. Some conclusions and future perspectives are given in the last section. Throughout this paper, we let $\mathbb{C}$, $\mathbb{R}$, $\mathbb{R}^+$, and $\mathbb{N}^+$ stand for the set of complex numbers, the set of real numbers, the set of positive real numbers, and the set of positive integers, respectively.

2. Preliminaries

In this section, we briefly review some basic definitions and facts concerning the k-hypergeometric series and the ordinary differential equation. Some surveys and literature for k-hypergeometric series and the k-hypergeometric differential equation can be found in Díaz et al. [30,31], Krasniqi [32,33], and Mubeen et al. [38,39].

Definition 1. *Assume that $x \in \mathbb{C}$, $k \in \mathbb{R}^+$ and $n \in \mathbb{N}^+$, then the Pochhammer k-symbol $(x)_{n,k}$ is defined by:*

$$(x)_{n,k} = x(x+k)(x+2k)\ldots[x+(n-1)k]. \tag{5}$$

In particular, we denote $(x)_{0,k} \equiv 1$. Therefore, we have the following facts:

(i) $(x)_{n+1,k} = (x+nk)(x)_{n,k}$.

(ii) $(1)_{n,1} = n!$; $(\frac{1}{2})_{n,1} = \dfrac{(2n-1)!!}{2^n}$; $(\frac{3}{2})_{n,1} = \dfrac{(2n+1)!!}{2^n}$.

(iii) $(x)_{n,1} = \dfrac{\Gamma(x+n)}{\Gamma(x)}$, where $\Gamma(x)$ is the Gamma function defined by $\int_0^\infty e^{-t}t^{x-1}dt$.

(iv) $(1)_{n,2} = (2n-1)!!$; $(2)_{n,2} = (2n)!!$; $(3)_{n,2} = (2n+1)!!$; $(4)_{n,2} = \dfrac{(2n+2)!!}{2}$.

Definition 2. *Assume that $a, b, c \in \mathbb{C}$, $k \in \mathbb{R}^+$ and $n \in \mathbb{N}^+$, then the k-hypergeometric series with three parameters a, b, and c is defined as:*

$$_2F_{1,k}[(a,k),(b,k);(c,k);z] = \sum_{n=0}^{\infty} \dfrac{(a)_{n,k}(b)_{n,k}}{(c)_{n,k}n!} z^n, \tag{6}$$

where $c \neq 0, -1, -2, -3, \ldots$ and $z \in \mathbb{C}$.

Definition 3. *Assume that $Y_0(z)$, $Y_1(z)$, and $Y_2(z)$ are three functions of z. Let a second-order ordinary differential equation be written in the following form:*

$$Y_2(z)\dfrac{d^2y}{dz^2} + Y_1(z)\dfrac{dy}{dz} + Y_0(z) = 0. \tag{7}$$

Then, the method about finding an infinite series solution of Equation (7) is called the Frobenius method.

Definition 4. *For Equation (7), let its coefficient $Y_2(z)$ satisfy $Y_2(z_0) = 0$ about the point $z_0 \in \mathcal{D} \subseteq \mathbb{C}$. Further, if this coefficient $Y_2(z)$ is holomorphic in a deleted neighborhood $\{z|\ 0 < |z - z_0| < \varepsilon\}$ for some $\varepsilon > 0$ and is meromorphic (not all holomorphic) in a neighborhood $\{z|\ |z - z_0| < \varepsilon\}$, then the point z_0 is called a singular point of Equation (7).*

Definition 5. *For the Equation (7), if the coefficient:*

$$Y_2(z) = (z - z_0)^i h(z)$$

is holomorphic at the point z_0, then the singular point z_0 of Equation (7) is said to be regular.

Dividing both sides of this Equation (7) by $Y_2(z)$ gives a differential equation of the following form:

$$\frac{d^2y}{dz^2} + \frac{Y_1(z)}{Y_2(z)}\frac{dy}{dz} + \frac{Y_0(z)}{Y_2(z)} = 0. \qquad (8)$$

As we know, if either $\frac{Y_0(z)}{Y_2(z)}$ or $\frac{Y_1(z)}{Y_2(z)}$ is not analytic at any regular singular point z_0, then Equation (8) cannot be solvable with the regular power series method. However, the method of Frobenius enables us to gain a power series solution of the differential equation defined by Equation (8), provided that both $\frac{Y_0(z)}{Y_2(z)}$ and $\frac{Y_1(z)}{Y_2(z)}$ are themselves analytic at z_0 or they are analytic elsewhere and their limits exist at z_0.

3. The Solutions of Non-Homogeneous k-Hypergeometric Equations

In this section, by means of the Frobenius method, we expound upon the series solution of the second-order non-homogeneous k-hypergeometric ordinary differential equation defined by Equation (4). Before presenting the main results, in order to judge whether a series is convergent or not, we usually need to apply the following criterion.

Lemma 1 (The d'Alembert test). *If the series $\sum_{n=0}^{\infty} u_n$ satisfies the following condition:*

$$\lim_{n \to \infty} \left| \frac{u_{n+1}}{u_n} \right| < 1 \quad (resp. > 1),$$

then this series $\sum_{n=0}^{\infty} u_n$ converges (resp. diverges).

Proof. Let us recall the following fact: The geometric series:

$$\sum_{n=0}^{\infty} q^n \quad (q > 0),$$

converges (resp. diverges) if $q < 1$ (resp. $q > 1$). Then, the proof of Lemma 1 is a simple series exercise. □

Theorem 1. *Suppose that $k \in \mathbb{R}^+$ and all a, b, c belong to $\mathbb{R}$. Let, in addition, c and $2k - c$ be neither zero, nor negative integers. Then, the homogeneous k-hypergeometric ordinary differential Equation (3) can have a general solution in the following form:*

$$y(z) = A \,_2F_{1,k}[(a,k),(b,k);(c,k);z] + B\, z^{1-\frac{c}{k}} \,_2F_{1,k}[(a+k-c,k),(b+k-c,k);(2k-c,k);z] \quad (9)$$

for $|z| < 1/k$, where A and B are two constants in $\mathbb{C}$.

Proof. Assume that:

$$y(z) = z^g \sum_{i=0}^{\infty} u_i z^i \qquad (10)$$

is any solution of the homogeneous ordinary differential Equation (3) with $u_0 \neq 0$. Then, differentiating Equation (10) directly, one has:

$$y'(z) = z^g \sum_{i=0}^{\infty} u_i(i+g)z^{i-1} \qquad (11)$$

and:

$$y''(z) = z^g \sum_{i=0}^{\infty} u_i(i+g)(i+g-1)z^{i-2}. \qquad (12)$$

Substituting Equations (11) and (12) into the k-hypergeometric differential Equation (3), we have:

$$kz^g \sum_{i=0}^{\infty} u_i(i+g)(i+g-1)z^{i-1} - k^2 z^g \sum_{i=0}^{\infty} u_i(i+g)(i+g-1)z^i \qquad (13)$$

$$+ cz^g \sum_{i=0}^{\infty} u_i(i+g)z^{i-1} - k(a+b+k)z^g \sum_{i=0}^{\infty} u_i(i+g)z^i - abz^g \sum_{i=0}^{\infty} u_i z^i = 0,$$

or, equivalently,

$$u_0 g(k(g-1) + c)z^{g-1} + kz^g \sum_{i=1}^{\infty} u_i(i+g)(i+g-1)z^{i-1} \qquad (14)$$

$$- k^2 z^g \sum_{i=1}^{\infty} u_{i-1}(i+g-1)(i+g-2)z^{i-1} + cz^g \sum_{i=1}^{\infty} u_i(i+g)z^{i-1}$$

$$- k(a+b+k)z^g \sum_{i=1}^{\infty} u_{i-1}(i+g-1)z^{i-1} - abz^g \sum_{i=1}^{\infty} u_{i-1}z^{i-1} = 0.$$

By comparing the coefficients on both sides of Equation (14), one can obtain the indicial equation:

$$g[k(g-1) + c] = 0 \qquad (15)$$

and the difference equation:

$$(g+i+1)[k(g+i) + c]u_{i+1} = [k(g+i) + a][k(g+i) + b]u_i, \qquad (16)$$

for $i = 0, 1, 2, \ldots$.

Solving the above indicial Equation (15) for g gives:

$$g = 0 \text{ and } g = 1 - \frac{c}{k}. \qquad (17)$$

Next, we discuss the solution of Equation (3) in two cases.

- Case 1: $g = 0$.

From Equation (16), we have the solution of Equation (3):

$$y_1(z) = u_0 \, _2F_{1,k}[(a,k),(b,k);(c,k);z], \qquad (18)$$

provided that c is not zero or a negative integer.

- Case 2: $g = 1 - \frac{c}{k}$.

In a similar manner, from Equation (16), we get the difference equation as follows:

$$(1+i)(2k-c+ki)u_{i+1} = (a+k-c+ki)(b+k-c+ki)u_i. \qquad (19)$$

Therefore, it follows that the other solution of Equation (3) is written as below:

$$y_2(z) = u_0 z^{1-\frac{c}{k}} \, _2F_{1,k}[(a+k-c,k),(b+k-c,k);(2k-c,k);z], \qquad (20)$$

provided that $2k - c$ is not a negative integer or zero.

Furthermore, from Equations (18) and (20), let us consider the radius of convergence of the series:

$$\sum_{i=0}^{\infty} v_i = {}_2F_{1,k}[(a,k),(b,k);(c,k);z]$$

and:

$$\sum_{i=0}^{\infty} w_i = {}_2F_{1,k}[(a+k-c,k),(b+k-c,k);(2k-c,k);z].$$

Referring to Lemma 1, we verify that:

$$\lim_{i\to\infty}\left|\frac{v_{i+1}}{v_i}\right| = \lim_{i\to\infty}\left|\frac{(a+ki)(b+ki)}{(c+ki)(1+i)}z\right| = |kz| < 1$$

and:

$$\lim_{i\to\infty}\left|\frac{w_{i+1}}{w_i}\right| = \lim_{i\to\infty}\left|\frac{(a+k-c+ki)(b+k-c+ki)}{(2k-c+ki)(1+i)}z\right| = |kz| < 1,$$

which imply that the series $\sum_{i=0}^{\infty} v_i$ and $\sum_{i=0}^{\infty} w_i$ have the same radius of convergence $\frac{1}{k}$.

Therefore, the general solution of the k-hypergeometric differential Equation (3) can be written as:

$$\begin{aligned} y(z) =& \alpha y_1(z) + \beta y_2(z) = A\ {}_2F_{1,k}[(a,k),(b,k);(c,k);z] \\ &+ B\ z^{1-\frac{c}{k}}\ {}_2F_{1,k}[(a+k-c,k),(b+k-c,k);(2k-c,k);z] \end{aligned} \quad (21)$$

for $|z| < 1/k$, where α, β, A, and B are four constants in $\mathbb{C}$.

Therefore, we have completed the proof of Theorem 1. □

Next, when the function $f(z)$ is a polynomial, that is:

$$f(z) = \sum_{i=0}^{m} d_i z^i, (m = 0,1,2,\ldots), \quad (22)$$

where $d_i, i = 0,1,2,\ldots,m$, are real or complex constants, we consider the solution of the non-homogeneous k-hypergeometric ordinary differential equation. The following theorem gives the particular solution and general solution of Equation (4).

Theorem 2. *Suppose that $k \in \mathbb{R}^+$ and all a, b, c belong to $\mathbb{R}$. Let, in addition, c and $2k - c$ be neither zero, nor negative integers. Then, the non-homogeneous k-hypergeometric ordinary differential Equation (4) can have a particular solution in the following form:*

$$\bar{y}(z) = -\sum_{j=0}^{m}\left[\frac{(a)_{j,k}(b)_{j,k}}{j!(c)_{j,k}}\sum_{l=j}^{m}\frac{l!(c)_{l,k}}{(a)_{l+1,k}(b)_{l+1,k}}d_l\right]z^j. \quad (23)$$

Therefore, a general solution of Equation (4) can be written as:

$$\begin{aligned} y(z) =& A\ {}_2F_{1,k}[(a,k),(b,k);(c,k);z] \\ &+ B\ z^{1-\frac{c}{k}}\ {}_2F_{1,k}[(a+k-c,k),(b+k-c,k);(2k-c,k);z] \\ &- \sum_{j=0}^{m}\left[\frac{(a)_{j,k}(b)_{j,k}}{j!(c)_{j,k}}\sum_{l=j}^{m}\frac{l!(c)_{l,k}}{(a)_{l+1,k}(b)_{l+1,k}}d_l\right]z^j \end{aligned} \quad (24)$$

for $|z| < 1/k$, where A and B are two constants in $\mathbb{C}$.

Proof. Let us assume that:
$$\bar{y}(z) = \sum_{j=0}^{m} s_j z^j \qquad (25)$$

is a particular solution of the non-homogeneous k-hypergeometric ordinary differential Equation (4), where $s_j, j = 0, 1, 2, \ldots, m$, are undetermined coefficients. Differentiating Equation (25), then we have:

$$\bar{y}'(z) = \sum_{j=0}^{m-1} (j+1) s_{j+1} z^j \qquad (26)$$

and:

$$\bar{y}''(z) = \sum_{j=0}^{m-2} (j+2)(j+1) s_{j+2} z^j. \qquad (27)$$

Plugging Equations (26) and (27) into Equation (4) yields:

$$kz(1-kz) \sum_{j=0}^{m-2} (j+2)(j+1) s_{j+2} z^j + [c - (k+a+b)kz] \sum_{j=0}^{m-1} (j+1) s_{j+1} z^j - ab \sum_{j=0}^{m} s_j z^j = \sum_{i=0}^{m} d_i z^i, \qquad (28)$$

and it follows that:

$$cs_1 - abs_0 + [2 \cdot 1 \cdot ks_2 + 2cs_2 - (k+a+b)ks_1 - abs_1]z \qquad (29)$$
$$+ [-2 \cdot 1 \cdot k^2 s_2 + 3 \cdot 2 \cdot ks_3 - 2k(k+a+b)s_2 + 3cs_3 - abs_2]z^2$$
$$+ \ldots$$
$$+ [m(m-1)ks_m - (m-1)(m-2)k^2 s_{m-1} + mcs_m$$
$$\qquad - (m-1)(k+a+b)ks_{m-1} - abs_{m-1}]z^{m-1}$$
$$+ [-m(m-1)k^2 s_m - mk(k+a+b)s_m - abs_m]z^m$$
$$= d_0 + d_1 z + d_2 z^2 + \ldots d_m z^m.$$

Matching the coefficients on both sides of Equation (29) gives:

$$\begin{cases} cs_1 - abs_0 = d_0, \\ 2 \cdot 1 \cdot ks_2 + 2cs_2 - (k+a+b)ks_1 - abs_1 = d_1, \\ \ldots \\ m(m-1)ks_m - (m-1)(m-2)k^2 s_{m-1} + mcs_m \\ \qquad - (m-1)(k+a+b)ks_{m-1} - abs_{m-1} = d_{m-1}, \\ -m(m-1)k^2 s_m - mk(k+a+b)s_m - abs_m = d_m. \end{cases} \qquad (30)$$

Thus, Equation (30) implies that:

$$s_m = -\frac{d_m}{m(m-1)k^2 + mk(k+a+b) + ab} = -\frac{d_m}{(a+km)(b+km)}$$

and:

$$\begin{aligned}s_{m-1} &= \frac{m(m-1)ks_m + mcs_m - d_{m-1}}{k^2(m-1)(m-2) + k(m-1)(k+a+b) + ab} \\
&= \frac{m[c+(m-1)k]}{[a+(m-1)k][b+(m-1)k]} s_m - \frac{d_{m-1}}{[a+(m-1)k][b+(m-1)k]} \\
&= \frac{m[c+(m-1)k]}{[a+(m-1)k][b+(m-1)k]} \cdot \frac{-d_m}{(a+km)(b+km)} - \frac{d_{m-1}}{[a+(m-1)k][b+(m-1)k]} \\
&= -\frac{(a)_{m-1,k}(b)_{m-1,k}}{(m-1)!(c)_{m-1,k}} \frac{m!(c)_{m,k}}{(a)_{m+1,k}(b)_{m+1,k}} d_m - \frac{d_{m-1}}{[a+(m-1)k][b+(m-1)k]} \\
&= -\frac{(a)_{m-1,k}(b)_{m-1,k}}{(m-1)!(c)_{m-1,k}} \left[\frac{m!(c)_{m,k}}{(a)_{m+1,k}(b)_{m+1,k}} d_m + \frac{(m-1)!(c)_{m-1,k}}{(a)_{m,k}(b)_{m,k}} d_{m-1} \right] \\
&= -\frac{(a)_{m-1,k}(b)_{m-1,k}}{(m-1)!(c)_{m-1,k}} \sum_{l=m-1}^{m} \frac{l!(c)_{l,k}}{(a)_{l+1,k}(b)_{l+1,k}} d_l.\end{aligned}$$

Consequently, these coefficients of the particular solution (25) are:

$$s_j = -\frac{(a)_{j,k}(b)_{j,k}}{j!(c)_{j,k}} \sum_{l=j}^{m} \frac{l!(c)_{l,k}}{(a)_{l+1,k}(b)_{l+1,k}} d_l \tag{31}$$

for $j = 0, 1, 2, \ldots, m$.

Replacing Equation (25) with Equation (31), we obtain a particular solution of the equation in the following form:

$$\bar{y}(z) = -\sum_{j=0}^{m} \left[\frac{(a)_{j,k}(b)_{j,k}}{j!(c)_{j,k}} \sum_{l=j}^{m} \frac{l!(c)_{l,k}}{(a)_{l+1,k}(b)_{l+1,k}} d_l \right] z^j. \tag{32}$$

From the above Equation (32), we get a general solution of Equation (4), as shown in Equation (24). We have shown Theorem 2. □

4. Examples

Example 1. *Find the solution to the following differential equation:*

$$z(1-z)\frac{d^2y}{dz^2} + (\frac{1}{2} - 3z)\frac{dy}{dz} - y = 1 + 2z^2 \tag{33}$$

for $|z| < 1$.

From Equation (33), it is easy to see that $a = b = 1$, $c = \frac{1}{2}$, $k = 1$, $m = 2$, $d_0 = 1$, $d_1 = 0$, and $d_2 = 2$. By Equation (24) in Theorem 2, then we have:

$$_2F_{1,k}[(a,k),(b,k);(c,k);z] = {_2F_{1,1}}[(1,1),(1,1);(\frac{1}{2},1);z] = \sum_{i=0}^{\infty} \frac{i! \cdot i!}{(\frac{1}{2})_{i,1} i!} z^i \tag{34}$$

$$= \sum_{i=0}^{\infty} \frac{i!}{(\frac{1}{2})_{i,1}} z^i = 1 + \sum_{i=1}^{\infty} \frac{2^i i!}{(2i-1)!!} z^i,$$

$$_2F_{1,k}[(a+k-c,k),(b+k-c,k);(2k-c,k);z] = {_2F_{1,1}}[(\frac{3}{2},1),(\frac{3}{2},1);(\frac{3}{2},1);z] \tag{35}$$

$$= \sum_{i=0}^{\infty} \frac{(\frac{3}{2})_{i,1}}{i!} z^i = \sum_{i=0}^{\infty} \frac{(2i+1)!!}{2^i i!} z^i$$

and:
$$\bar{y}(z) = -\sum_{i=0}^{2}\left[\frac{(1)_{i,1}(1)_{i,1}}{i!(\frac{1}{2})_{i,1}}\sum_{l=i}^{2}\frac{l!(\frac{1}{2})_{l,1}d_l}{(1)_{l+1,1}(1)_{l+1,1}}\right]z^i = -\left(\frac{13}{12}+\frac{1}{6}z+\frac{2}{9}z^2\right). \quad (36)$$

Combining Equations (34)–(36), we obtain the general solution of Equation (33) as below:

$$y = A\ {}_2F_{1,1}[(1,1),(1,1);(\tfrac{1}{2},1);z] + B\ \sqrt{z}\ {}_2F_{1,1}[(\tfrac{3}{2},1),(\tfrac{3}{2},1);(\tfrac{3}{2},1);z] + \bar{y}(z) \quad (37)$$

$$= A\left(1 + \sum_{i=1}^{\infty}\frac{2^i i!}{(2i-1)!!}z^i\right) + B\sqrt{z}\left(\sum_{i=0}^{\infty}\frac{(2i+1)!!}{2^i i!}z^i\right) - \left(\frac{13}{12}+\frac{1}{6}z+\frac{2}{9}z^2\right),$$

where A and B are two constants.

Example 2. *Find the solution to the following differential equation:*

$$2z(1-2z)\frac{d^2y}{dz^2} + (1-14z)\frac{dy}{dz} - 6y = 1 + 2z + 3z^2 \quad (38)$$

for $|z| < \tfrac{1}{2}$.

From Equation (38), it is clear that $k = 2, c = 1, m = 2$, and $d_0 = 1, d_1 = 2, d_2 = 3$. Then, let us take $a = 3, b = 2$. According to Equation (24) in Theorem 2, we obtain:

$$_2F_{1,k}[(a,k),(b,k);(c,k);z] = {}_2F_{1,2}[(3,2),(2,2);(1,2);z] \quad (39)$$

$$= \sum_{i=0}^{\infty}\frac{(2i+1)!!\ (2i)!!}{(2i-1)!!\ i!}z^i = \sum_{i=0}^{\infty}2^i(2i+1)z^i,$$

$$_2F_{1,k}[(a+k-c,k),(b+k-c,k);(2k-c,k);z] = {}_2F_{1,2}[(4,2),(3,2);(3,2);z] \quad (40)$$

$$= \sum_{i=0}^{\infty}\frac{(4)_{i,2}}{i!}z^i = \sum_{i=0}^{\infty}2^i(i+1)z^i$$

and:

$$\bar{y}(z) = -\sum_{i=0}^{2}\left[\frac{(3)_{i,2}(2)_{i,2}}{i!(1)_{i,2}}\sum_{l=i}^{2}\frac{l!(1)_{l,2}d_l}{(3)_{l+1,2}(2)_{l+1,2}}\right]z^i = -\left(\frac{157}{840}+\frac{17}{140}z+\frac{1}{14}z^2\right). \quad (41)$$

Substituting Equations (39)–(41) into Equation (24) gives the general solution of Equation (38) as follows:

$$y = A\ {}_2F_{1,2}[(3,2),(2,2);(1,2);z] + B\ \sqrt{z}\ {}_2F_{1,2}[(4,2),(3,2);(3,2);z] + \bar{y}(z) \quad (42)$$

$$= A\sum_{i=0}^{\infty}2^i(2i+1)z^i + B\sqrt{z}\left(\sum_{i=0}^{\infty}2^i(i+1)z^i\right) - \frac{157}{840} - \frac{17}{140}z - \frac{1}{14}z^2,$$

where A and B are two constants.

5. Conclusions

When $|z| < 1/k$ and $f(z) = \sum\limits_{i=0}^{m} d_i z^i, d_i \in \mathbb{C}, i = 0, 1, 2, \ldots, m$, in this paper, we present a formula of the general solution of the non-homogeneous k-hypergeometric ordinary differential Equation (4), provided that $a, b, c \in \mathbb{R}$ with both c and $2k - c$ neither zero, nor negative integers. The solutions of this type of equations are denoted by in the form of k-hypergeometric series, and it is convenient that we can make out the corresponding computer program and put it into calculation. When $f(z)$ is not a polynomial, it is a fascinating question to derive the particular or general series solutions for the non-homogeneous k-hypergeometric ordinary differential Equation (1). It would be interesting to have more research about this case.

Author Contributions: The contributions of all of the authors have been similar. All of them have worked together to develop the present manuscript.

Funding: This research was funded by the Project of University-Industry Cooperation of Ministry of Education of P.R. China (Grant Nos. CX2015ZG23, 201702023030), the Project of Quality Curriculums of Education Department of Anhui Province (Grant No.2016gxx087), the Natural Science Key Foundation of Education Department of Anhui Province (Grant No.KJ2013A183), the Project of Leading Talent Introduction and Cultivation in Colleges and Universities of Education Department of Anhui Province (Grant No. gxfxZD2016270) and the Incubation Project of the National Scientific Research Foundation of Bengbu University (Grant No. 2018GJPY04).

Acknowledgments: The authors are thankful to the anonymous reviewers for their valuable comments.

Conflicts of Interest: The authors declare no conflict of interest.

References

1. Euler, L. *Institutionum Calculi Integralis*; Impensis Academiae Imperialis Scientiarum: Petropolis, Brazil, 1769.
2. Coddington, E.A.; Levinson, N. *Theory of Ordinary Differential Equations*; McGraw–Hill: New York, NY, USA, 1955.
3. Teschl, G. *Ordinary Differential Equations and Dynamical Systems*; American Mathematical Society: Providence, RI, USA, 2012.
4. Kurilin, B.I. Solution of the general Riccati equation with the aid of continued fractions. *Radiophys. Quantum Electron.* **1968**, *11*, 640–641. [CrossRef]
5. Ku, Y.H. Solution of the riccati equation by continued fractions. *J. Frankl. Inst.* **1972**, *293*, 59–65. [CrossRef]
6. Arnold, C. Formal continued fractions solutions of the generalized second order Riccati equations, applications. *Numer. Algorithms* **1997**, *15*, 111–134. [CrossRef]
7. Elizalde, E.;Odintsov, S.D.;Romeo, A.;Bytsenko, A.; Zerbini, S. *Zeta Regularization Techniques with Applications*; World Scientific Publishing Company: Singapore, 1994.
8. Elizalde, E. Analysis of an inhomogeneous generalized Epstein-Hurwitz zeta function with physical applications. *J. Math. Phys.* **1994**, *35* 6100–6122. [CrossRef]
9. Bordag, M.; Elizalde, E.; Kirsten, K. Heat kernel coefficients of the Laplace operator on the D-dimensional ball. *J. Math. Phys.*, **1996**, *37*, 895–916. [CrossRef]
10. Elizalde, E. *Ten Physical Applications of Spectral Zeta Functions*; Lecture Notes in Physics Book Series; Springer-Verlag: Berlin/Heidelberg, Germany, 2012.
11. Wallis, J. *Arithmetica Infinitorum (Latin)*; Originally published (1656); English translation: The Arithmetic of Infinitesimals by J. A. Stedall; Springer-Verlag: New York, NY, USA, 2004. [CrossRef]
12. Gauss, C.F. Disquisitiones generales circa seriem infinitam $1 + \frac{\alpha\beta}{1\cdot\gamma}x + \frac{\alpha(1+\alpha)\beta(1+\beta)}{1\cdot 2\cdot\gamma(1+\gamma)}xx + etc.$ pars prior. *Comm. Soc. Regiae Sci. Gottingensis Rec.* **1812**, *2*, 123–162.
13. Kummer, E.E. Über die hypergeometrische Reihe. *J. Die Reine Angew. Math.* **1836**, *15*, 39–83. [CrossRef]
14. Riemann, B. Beiträge zur Theorie der durch die Gauss'sche Reihe $F(\alpha, \beta, \gamma, x)$ darstellbaren Functionen. *Aus dem Sieben. Band Abh. Königlichen Gesellschaft Wiss. zu Göttingen* **1857**, *7*, 3–22.
15. Bailey, W.N. Transformations of generalized hypergeometric series. *Proc. Lond. Math. Soc.* **1929**. [CrossRef]
16. Bailey, W.N. *Generalized Hypergeometric Series*; Cambridge University Press: Cambridge, UK, 1935.
17. Chaundy, T.W. An extension of hypergeometric functions (I). *Q. J. Math.* **1943**, *14*, 55–78. [CrossRef]
18. Srivastava, H.M.; Karlsson, P.W. *Multiple Gaussian Hypergeometric Series*; Halsted Press: New York, NY, USA, 1985.

19. Whittaker, E.T.; Watson, G.N. *A Course of Modern Analysis*, 4th ed.; Cambridge University Press: Cambridge, UK, 1996.
20. Beukers, F. Gauss' hypergeometric function. *Prog. Math.* **2002**, *228*, 77–86.
21. Gasper, G.; Rahman, M. *Basic Hypergeometric Series*, 2nd ed.; Cambridge University Press: Cambridge, UK, 2004.
22. Olde Daalhuis, A.B. Hyperterminants I. *J. Comput. Appl. Math.* **1996**, *76*, 255–264. [CrossRef]
23. Olde Daalhuis, A.B. Hyperterminants II. *J. Comput. Appl. Math.* **1998**, *89*, 87–95. [CrossRef]
24. Dwork, B.; Loeser, F. Hypergeometric series. *Jpn. J. Math.* **1993**, *19*, 81–129. [CrossRef]
25. Chu, W. Terminating hypergeometric $_2F_1(2)$-series. *Integral Transform. Spec. Funct.* **2011**, *22*, 91–96. [CrossRef]
26. Yilmazer, R.; Inc, M.; Tchier, F.; Baleanu, D. Particular solutions of the confluent hypergeometric differential equation by using the Nabla fractional calculus operator. *Entropy* **2016**, *18*, 49. [CrossRef]
27. Morita, T.; Sato, K.-I. Solution of inhomogeneous differential equations with polynomial coefficients in terms of the Green's function. *Mathematics* **2017**, *5*, 62. [CrossRef]
28. Abramov, S.A.; Ryabenko, A.A.; Khmelnov, D.E. Laurent, rational, and hypergeometric solutions of linear q-difference systems of arbitrary order with polynomial coefficients. *Progr. Comput. Softw.* **2018**, *44*, 120–130. [CrossRef]
29. Alfedeel, A.H.A.; Abebe, A.; Gubara, H.M. A generalized solution of Bianchi type-V models with time-dependent G and Λ. *Universe* **2018**, *4*, 83. [CrossRef]
30. Díaz, R.; Teruel, C. q, k-Generalized Gamma and Beta functions. *J. Nonlinear Math. Phys.* **2005**, *12*, 118–134. [CrossRef]
31. Díaz, R.; Pariguan, E. On hypergeometric functions and Pochhammer k-symbol. *Divulg. Mat.* **2007**, *15*, 179–192.
32. Krasniqi, V. A limit for the k-gamma and k-beta function. *Int. Math. Forum* **2010**, *5*, 1613–1617.
33. Krasniqi, V. Inequalities and monotonicity for the ration of k-gamma functions. *Sci. Magna* **2010**, *6*, 40–45.
34. Kokologiannaki, C.G. Properties and inequalities of generalized k-gamma, beta and zeta functions. *Int. J. Contemp. Math. Sci.* **2010**, *5*, 653–660.
35. Mubeen, S.; Habibullah, G.M. An integral representation of some k-hypergeometric functions. *Int. Math. Forum* **2012**, *7*, 203–207.
36. Mubeen, S.; Habibullah, G.M. k-fractional integrals and application. *Int. J. Contemp. Math. Sci.* **2012**, *7*, 89–94.
37. Mubeen, S. Solution of some integral equations involving conuent k-hypergeometricfunctions. *Appl. Math.* **2013**, *4*, 9–11. [CrossRef]
38. Mubeen, S.; Rehman, A. A Note on k-Gamma function and Pochhammer k-symbol. *J. Inf. Math. Sci.* **2014**, *6*, 93–107.
39. Mubeen, S.; Naz, M.A. Rehman, G. Rahman. Solutions of k-hypergeometric differential equations. *J. Appl. Math.* **2014**, *2014*, 1–13. [CrossRef]
40. Mubeen, S.; Iqbal, S. Some inequalities for the gamma k-function. *Adv. Inequal. Appl.* **2015**, *2015*, 1–9.
41. Rehman, A.; Mubeen, S.; Sadiq, N.; Shaheen, F. Some inequalities involving k-gamma and k-beta functions with applications. *J. Inequal. Appl.* **2014**, *2014*, 224. [CrossRef]
42. Rahman, G.; Arshad, M.; Mubeen, S. Some results on a generalized hypergeometric k-functions. *Bull. Math. Anal. Appl.* **2016**, *8*, 6–77.

© 2019 by the authors. Licensee MDPI, Basel, Switzerland. This article is an open access article distributed under the terms and conditions of the Creative Commons Attribution (CC BY) license (http://creativecommons.org/licenses/by/4.0/).

Article

Symplectic Effective Order Numerical Methods for Separable Hamiltonian Systems

Junaid Ahmad [1], Yousaf Habib [2,3,*], Shafiq ur Rehman [1], Azqa Arif [3], Saba Shafiq [3] and Muhammad Younas [4]

[1] Department of Mathematics, University of Engineering and Technology, Lahore 54890, Pakistan; junaidshaju@gmail.com (J.A.); srehman@uet.edu.pk (S.u.R.)
[2] Department of Mathematics, COMSATS University Islamabad, Lahore Campus, Lahore 45550, Pakistan
[3] School of Natural Sciences, Department of Mathematics, National University of Sciences and Technology, Islamabad 44000, Pakistan; azqaarif@gmail.com (A.A.); innocentravian@yahoo.com (S.S.)
[4] Business Administration Programme, Virtual University of Pakistan, Lahore 54000, Pakistan; younasnaz99@gmail.com
* Correspondence: yhabib@gmail.com; Tel.: +92-333-572-3457

Received: 7 January 2019; Accepted: 22 January 2019; Published: 28 January 2019

Abstract: A family of explicit symplectic partitioned Runge-Kutta methods are derived with effective order 3 for the numerical integration of separable Hamiltonian systems. The proposed explicit methods are more efficient than existing symplectic implicit Runge-Kutta methods. A selection of numerical experiments on separable Hamiltonian system confirming the efficiency of the approach is also provided with good energy conservation.

Keywords: effective order; partitioned runge-kutta methods; symplecticity; hamiltonian systems

---

1. Introduction

Over the last few decades, a lot of progress has been made in developing numerical methods for ordinary differential equations (ODEs), which can produce efficient, reliable and qualitatively correct numerical solutions by preserving some qualitative features of the exact solutions [1,2]. This field of research is called geometric numerical integration. In this paper, we are oriented to obtain numerical approximation of Hamiltonian differential equations of the form,

$$\frac{dp}{dt} = -\frac{\partial H}{\partial q}, \qquad \frac{dq}{dt} = \frac{\partial H}{\partial p}, \qquad (1)$$

where H is known as the Hamiltonian or the total energy of the system and q and p are generalized coordinates and generalized momenta respectively. The autonomous Hamiltonian systems have two important properties; one is that the total energy remains constant,

$$\frac{dH}{dt} = \frac{\partial H}{\partial p} \cdot \frac{\partial p}{\partial t} + \frac{\partial H}{\partial q} \cdot \frac{\partial q}{\partial t} = 0.$$

The second important property is that the phase flow is symplectic which imply that the motion along the phase curve retains the area of a bounded sub-domain in the phase space. We need such numerical methods which can mimic both properties of the Hamiltonian systems. For this we use symplectic numerical methods to solve system of Equation (1). The symplectic methods are numerically more efficient than non-symplectic methods for integration over long interval of time [3]. Among the class of one step symplectic methods, the implicit symplectic Runge–Kutta (RK) methods were developed and presented in [4].

For general explicit RK methods, it is known that up to order 4, number of stages required in a RK method are equal to the order of the method whereas, for order ≥ 5, number of stages become greater than the order of the method. Thus for example, an RK method of order 5 needs at least 6 stages and an RK method of order 6 requires at least 17 stages and so on. The more the number of stages are, the more costly the method is.

Butcher [5] tried to overcome this complexity of order barrier by presenting the idea of effective order. He implemented his idea on RK method of order 5 and was able to construct explicit RK methods of effective order 5 with just 5 internal stages [5]. Later on, the idea was extended to construct diagonally extended singly implicit RK methods for the numerical integration of stiff differential equations [6–9]. Butcher also used the effective order technique on symplectic RK methods for the numerical integration of Hamiltonian systems [10,11].

A RK method v has an effective order q if we have another RK method w, known as the starting method, such that $wv^n w^{-1}$ has an order q. The method w is used only once in the beginning followed by n iterations of the main method v and the method w^{-1}, known as the finishing method is used only once at the end.

For separable Hamiltonian systems, it is advantageous to solve some components of the differential-equation with one RK method and solve other components of differential equation with another RK method and collectively they are termed as partitioned Runge–Kutta (PRK) methods. We have extended the idea of Butcher to construct symplectic effective order PRK methods which are explicit in nature and hence are less costly than symplectic implicit RK methods. For the effective order PRK methods, we construct two main PRK methods, two starting and two finishing PRK methods such that the two starting methods are applied once at the beginning followed by n iterations of the main PRK methods and the two finishing methods are used at the end.

2. Algebraic Structure of PRK Methods

We are concerned with the numerical solution of separable Hamiltonian systems,

$$\begin{pmatrix} u \\ v \end{pmatrix}' = \begin{pmatrix} k(v) \\ r(u) \end{pmatrix}, \quad u(t_0) = u_0, \; v(t_0) = v_0, \; k, r : \mathbb{R}^N \to \mathbb{R}^N, \tag{2}$$

using two s-stages RK methods $M = [a \; b \; c]$ and $\tilde{M} = [\tilde{a} \; \tilde{b} \; \tilde{c}]$,

$$U_i = u_n + h \sum_{j=1}^{s-1} a_{ij} k(V_j), \qquad V_i = v_n + h \sum_{j=1}^{s} \tilde{a}_{ij} r(U_j), \qquad i = 1, 2, ..., s,$$

$$u_{n+1} = u_n + h \sum_{j=1}^{s} b_j k(V_j), \qquad v_{n+1} = v_n + h \sum_{j=1}^{s} \tilde{b}_j r(U_j),$$

where U_i and V_i are the stages for the u and v variables, b_i and $\tilde{b}_i$ are quadrature weights, c_i and $\tilde{c}_i$ are quadrature nodes, $A = (a_{ij})_{s \times s}$ and $\tilde{A} = (\tilde{a}_{ij})_{s \times s}$ are matrices of s-stage PRK methods M and $\tilde{M}$, respectively. Here M is an explicit RK method and $\tilde{M}$ is an implicit RK method. Similarly, we can define two PRK methods $S = [A \; B \; C]$ and $\tilde{S} = [\tilde{A} \; \tilde{B} \; \tilde{C}]$, termed as starting methods. The four PRK methods can be represented by Butcher tableaux as,

$$\begin{array}{c|c} c & a \\ \hline & b^T \end{array}, \quad \begin{array}{c|c} \tilde{c} & \tilde{a} \\ \hline & \tilde{b}^T \end{array}, \quad \begin{array}{c|c} C & A \\ \hline & B^T \end{array} \quad \text{and} \quad \begin{array}{c|c} \tilde{C} & \tilde{A} \\ \hline & \tilde{B}^T \end{array}.$$

We need to find the composed methods MS and $\widetilde{MS}$, which are given by the Butcher tableaux,

$$
\begin{array}{c|cc}
c & A & 0 \\
C + \sum_{i=1}^{s} b_i & b & A \\
\hline
& b^T & B^T
\end{array}
\quad , \quad
\begin{array}{c|cc}
\tilde{c} & \tilde{A} & 0 \\
\widetilde{C} + \sum_{i=1}^{s} \tilde{b}_i & \tilde{b} & \tilde{A} \\
\hline
& \tilde{b}^T & \tilde{B}^T
\end{array} . \quad (3)
$$

The composition asserts to carry out the calculations with starting methods S and $\widetilde{S}$ firstly and applying the PRK methods M and $\widetilde{M}$ to the output. To explain the composition (3), we consider four PRK methods with two stages each given as,

$$
\begin{array}{c|cc}
c_1 & a_{11} & a_{12} \\
c_2 & a_{21} & a_{22} \\
\hline
& b_1 & b_2
\end{array} \, , \quad
\begin{array}{c|cc}
\tilde{c}_1 & \tilde{a}_{11} & \tilde{a}_{12} \\
\tilde{c}_2 & \tilde{a}_{21} & \tilde{a}_{22} \\
\hline
& \tilde{b}_1 & \tilde{b}_2
\end{array} \, , \quad
\begin{array}{c|cc}
C_1 & A_{11} & A_{12} \\
C_2 & A_{21} & A_{22} \\
\hline
& B_1 & B_2
\end{array} \, , \quad
\begin{array}{c|cc}
\widetilde{C}_1 & \widetilde{A}_{11} & \widetilde{A}_{12} \\
\widetilde{C}_2 & \widetilde{A}_{21} & \widetilde{A}_{22} \\
\hline
& \widetilde{B}_1 & \widetilde{B}_2
\end{array} .
$$

Solving the differential equation $u' = k(v)$, we take the first step going from u_0 to u_1 using starting method S. We then take the second step going from u_1 to u_2 using main method M. The equations are given as,

$$U_1 = u_0 + a_{11} h\, k(V_1) + a_{12} h\, k(V_2), \quad \overline{U}_1 = u_1 + A_{11} h\, k(\overline{V}_1) + A_{12} h\, k(\overline{V}_2),$$
$$U_2 = u_0 + a_{21} h\, k(V_1) + a_{22} h\, k(V_2), \quad \overline{U}_2 = u_1 + A_{21} h\, k(\overline{V}_1) + A_{22} h\, k(\overline{V}_2),$$
$$u_1 = u_0 + b_1 h\, k(V_1) + b_2 h\, k(V_2), \quad u_2 = u_1 + B_1 h\, k(\overline{V}_1) + B_2 h\, k(\overline{V}_2).$$

The composition MS means that we are going from u_0 to u_2 using the composed method MS given as,

$$
\begin{aligned}
U_1 &= u_0 + a_{11} h\, k(V_1) + a_{12} h\, k(V_2), \\
U_2 &= u_0 + a_{21} h\, k(V_1) + a_{22} h\, k(V_2), \\
\overline{U}_1 &= u_0 + b_1 h\, k(V_1) + b_2 h\, k(V_2) + A_{11} h\, k(\overline{V}_1) + A_{12} h\, k(\overline{V}_2), \\
\overline{U}_2 &= u_0 + b_1 h\, k(V_1) + b_2 h\, k(V_2) + A_{21} h\, k(\overline{V}_1) + A_{22} h\, k(\overline{V}_2), \\
u_2 &= u_0 + b_1 h\, k(V_1) + b_2 h\, k(V_2) + B_1 h\, k(\overline{V}_1) + B_2 h\, k(\overline{V}_2).
\end{aligned} \quad (4)
$$

The Butcher table for the above composed PRK methods is as follows

$$
\begin{array}{c|cccc}
c_1 & a_{11} & a_{12} & 0 & 0 \\
c_2 & a_{21} & a_{22} & 0 & 0 \\
C_1 + \sum b_i & b_1 & b_2 & A_{11} & A_{12} \\
C_2 + \sum b_i & b_1 & b_2 & A_{21} & A_{22} \\
\hline
& b_1 & b_2 & B_1 & B_2
\end{array}
$$

Similarly, the composition $\widetilde{MS}$ can be obtained and is represented by the following Butcher table

$$
\begin{array}{c|cccc}
\tilde{c}_1 & \tilde{a}_{11} & \tilde{a}_{12} & 0 & 0 \\
\tilde{c}_2 & \tilde{a}_{21} & \tilde{a}_{22} & 0 & 0 \\
\widetilde{C}_1 + \sum \tilde{b}_i & \tilde{b}_1 & \tilde{b}_2 & \widetilde{A}_{11} & \widetilde{A}_{12} \\
\widetilde{C}_2 + \sum \tilde{b}_i & \tilde{b}_1 & \tilde{b}_2 & \widetilde{A}_{21} & \widetilde{A}_{22} \\
\hline
& \tilde{b}_1 & \tilde{b}_2 & \widetilde{B}_1 & \widetilde{B}_2
\end{array}
$$

A numerical scheme has order p, if after one iteration, the numerical solution matches with the Taylor's series of the exact solution to the extent that the remainder term has $O(h^{p+1})$. The comparison of the numerical solution with the Taylor's series of the exact solution provides order conditions which must be satisfied to obtain a particular order numerical method. Thus for order 3 method, we have the following order conditions:

$$\sum_i b_i = 1, \tag{5}$$

$$\sum_i \tilde{b}_i = 1, \tag{6}$$

$$\sum_i b_i \tilde{c}_i = \frac{1}{2}, \tag{7}$$

$$\sum_i \tilde{b}_i c_i = \frac{1}{2}, \tag{8}$$

$$\sum_i b_i \tilde{c}_i^2 = \frac{1}{3}, \tag{9}$$

$$\sum_i \tilde{b}_i c_i^2 = \frac{1}{3}, \tag{10}$$

$$\sum_{i,j} b_{i,j} \tilde{a}_{ij} c_j = \frac{1}{6}, \tag{11}$$

$$\sum_{i,j} \tilde{b}_i a_{ij} \tilde{c}_j = \frac{1}{6}. \tag{12}$$

3. Rooted Trees for PRK Methods and Order Conditions

There is a graph theoretical approach to study order conditions of RK methods due to Butcher [12]. We first study the basic concepts from graph theory and then relate the graphs to the order conditions of RK methods.

A graph with non-cyclic representation consisting of vertices and edges with one vertex acting as root is called as rooted tree. The rooted trees whose vertices are either black or white are known as bi-color rooted trees. The trees having black color root is termed as t, while the white rooted trees are represented by $\tilde{t}$. The order of a tree is total number of vertices in a tree. Whereas, the density $\gamma(t)$ is the product of number of vertices of a tree and their sub-trees, when we remove the root.

Example 1. *Consider a bi-color tree with order 5 and $\gamma(t) = 5 \times 4 \times 2 = 40$.*

The repeated differentiation of Equation (2) gives

$$\begin{aligned}
u^{(1)} &= k(v), & v^{(1)} &= r(u), \\
u^{(2)} &= \frac{\partial k}{\partial v} r, & v^{(2)} &= \frac{\partial r}{\partial u} k, \\
u^{(3)} &= \frac{\partial^2 k}{\partial v \, \partial v}(r,r) + \frac{\partial k}{\partial v} \frac{\partial r}{\partial u} k, & v^{(3)} &= \frac{\partial^2 r}{\partial u \, \partial u}(k,k) + \frac{\partial r}{\partial u} \frac{\partial k}{\partial v} r, \\
\vdots & & \vdots &
\end{aligned} \tag{13}$$

The terms on the right hand side of Equation (13) are called elementary differentials and can be represented graphically with the help of bi-color rooted trees. Thus k is represented by a black vertex and r is represented by a white vertex, whereas the differentiation is represented by an edge between two vertices. According to the nature of the differential Equation (2) in which k depends only on v and r depends only on u, we only consider trees in which a black vertex has a white child and vice versa [13]. Such trees are given in Tables 1 and 2. The elementary differentials can be represented by bi-color rooted trees as shown in Table 3.

Table 1. Composition rule for trees with black vertex root.

t_i	Tree	$(\beta\alpha)(t_i)$	$(E\beta)(t_i)$
t_1	●	$\beta_1 + \alpha_1$	$\beta_1 + 1$
t_2		$\beta_2 + \tilde{\beta}_1\alpha_1 + \alpha_2$	$\beta_2 + \beta_1 + \frac{1}{2}$
t_3		$\beta_3 + \tilde{\beta}_1^2\alpha_1 + 2\tilde{\beta}_1\alpha_2 + \alpha_3$	$\beta_3 + 2\beta_2 + \beta_1 + \frac{1}{3}$
t_4		$\beta_4 + \beta_1\alpha_2 + \tilde{\beta}_2\alpha_1 + \alpha_4$	$\beta_4 + \beta_2 + \frac{1}{2}\beta_1 + \frac{1}{6}$

Table 2. Composition rule for trees with white vertex root.

$\tilde{t}_i$	Tree	$(\tilde{\beta}\tilde{\alpha})(t_i)$	$(E\tilde{\beta})(t_i)$
$\tilde{t}_1$	○	$\tilde{\beta}_1 + \tilde{\alpha}_1$	$\tilde{\beta}_1 + 1$
$\tilde{t}_2$		$\tilde{\beta}_2 + \beta_1\tilde{\alpha}_1 + \tilde{\alpha}_2$	$\tilde{\beta}_2 + \tilde{\beta}_1 + \frac{1}{2}$
$\tilde{t}_3$		$\tilde{\beta}_3 + \beta_1^2\tilde{\alpha}_1 + 2\beta_1\tilde{\alpha}_2 + \tilde{\alpha}_3$	$\tilde{\beta}_3 + 2\tilde{\beta}_2 + \tilde{\beta}_1 + \frac{1}{3}$
$\tilde{t}_4$		$\tilde{\beta}_4 + \tilde{\beta}_1\tilde{\alpha}_2 + \beta_2\tilde{\alpha}_1 + \tilde{\alpha}_4$	$\tilde{\beta}_4 + \tilde{\beta}_2 + \frac{1}{2}\tilde{\beta}_1 + \frac{1}{6}$

Table 3. Elementary differentials and elementary weights of bi-color rooted trees.

t	Elementary Differentials	$\Phi(t)$	$\tilde{t}$	Elementary Differentials	$\Phi(\tilde{t})$
●	k	b_i	○	r	$\tilde{b}_i$
	$\frac{\partial k}{\partial v} r$	$b_i \tilde{c}_i$		$\frac{\partial r}{\partial u} k$	$\tilde{b}_i c_i$
	$\frac{\partial^2 k}{\partial v \partial v}(r,r)$	$b_i \tilde{c}_i^2$		$\frac{\partial^2 r}{\partial u \partial u}(k,k)$	$\tilde{b}_i c_i^2$
	$\frac{\partial k}{\partial v}\frac{\partial r}{\partial u} k$	$b_i \tilde{a}_{ij} c_j$		$\frac{\partial r}{\partial u}\frac{\partial k}{\partial v} r$	$\tilde{b}_i a_{ij} \tilde{c}_j$

The terms on left hand side from Equations (5)–(12) are called elementary weights $\Phi(t)$ and $\Phi(\tilde{t})$. The elementary weights are nonlinear expressions of the coefficients of PRK methods and can be related to bi-color rooted trees as shown in Table 3 [12,13]. A PRK method is of order p iff

$$\Phi(t) = \frac{1}{\gamma(t)}, \quad \text{and} \quad \Phi(\tilde{t}) = \frac{1}{\gamma(\tilde{t})},$$

for all bi-color rooted trees t and $\tilde{t}$ of order less than or equal to p [13].

Let the elementary weight functions of the methods M and $\tilde{M}$ are α and $\tilde{\alpha}$, respectively such that α maps trees t and $\tilde{\alpha}$ maps trees $\tilde{t}$ to expression in terms of method coefficients as follows

$$\alpha(\:\vcenter{\hbox{\scalebox{0.6}{Y}}}\:) = \sum b_i \tilde{c}_i^2, \qquad \tilde{\alpha}(\:\vcenter{\hbox{\scalebox{0.6}{Y}}}\:) = \sum \tilde{b}_i c_i^2.$$

Similarly, we can define elementary weight functions β and $\tilde{\beta}$ for the starting methods S and $\tilde{S}$, respectively. The composition of PRK methods in terms of their elementary weights is $\beta\alpha$ and $\tilde{\beta}\tilde{\alpha}$ given as

$$\beta\alpha(\:\vcenter{\hbox{\scalebox{0.6}{Y}}}\:) = \beta_3 + 2\tilde{\beta}_1\alpha_2 + \tilde{\beta}_1^2\alpha_1 + \alpha_3, \tag{14}$$

where the right hand side of Equation (14) contains terms from Table 4 which has tree u as a sub-tree of tree t and $t \setminus u$ is the remaining tree when u is removed from t.

Table 4. Calculation for the term $\beta\alpha(t_3)$.

Tree	No Cut	First Cut	Second Cut	Third Cut	All Cuts
t	Y	Y	Y	Y Y	Y
u		↑	↑	•	Y
$t \setminus u$	Y	∘	∘	∘ ∘	
term	β_3	$\tilde{\beta}_1\alpha_2$	$\tilde{\beta}_1\alpha_3$	$\tilde{\beta}_1^2\alpha_1$	α_3

The order condition related to the tree for the composed method (4) is

$$\begin{aligned} b_i\tilde{c}_i^2 + B_i(\tilde{C}_i + \tilde{b}_i)^2 &= b_i\tilde{c}_i^2 + 2\tilde{b}_iB_i\tilde{C}_i + \tilde{b}_i^2 B_i + B_i\tilde{C}_i^2, \\ &= \beta_3 + 2\tilde{\beta}_1\alpha_2 + \tilde{\beta}_1^2\alpha_1 + \alpha_3, \\ &= \beta\alpha(\:\vcenter{\hbox{\scalebox{0.6}{Y}}}\:). \end{aligned}$$

4. Effective Order PRK Methods

Let M and S be two RK methods together with an inverse method S^{-1}. The effective order q means that the composition SMS^{-1} has order q [6]. For PRK methods, we are interested to construct methods $M, \tilde{M}, S, \tilde{S}$ together with the inverse methods S^{-1} and $\tilde{S}^{-1}$ such that the compositions SMS^{-1} and $\tilde{S}\tilde{M}\tilde{S}^{-1}$ have the effective order q which implies $\beta\alpha(t) = E\beta(t)$ and $\tilde{\beta}\tilde{\alpha}(\tilde{t}) = E\tilde{\beta}(\tilde{t})$ for all trees of order q [12], where E is the exact flow for which corresponding order conditions are satisfied. The terms are given in Tables 1 and 2.

A method of effective order 3 is obtained by comparing two columns of Tables 1 and 2 each for trees up to order 3. Thus we have

$$\alpha_1 = 1, \tag{15}$$
$$\tilde{\alpha}_1 = 1, \tag{16}$$
$$\alpha_2 = \frac{1}{2}, \tag{17}$$
$$\tilde{\alpha}_2 = \frac{1}{2}, \tag{18}$$
$$\alpha_3 = \frac{1}{3} + 2\beta_2, \tag{19}$$
$$\tilde{\alpha}_3 = \frac{1}{3} + 2\tilde{\beta}_2, \tag{20}$$
$$\alpha_4 = \frac{1}{6} + \beta_2 - \tilde{\beta}_2, \tag{21}$$
$$\tilde{\alpha}_4 = \frac{1}{6} + \tilde{\beta}_2 - \beta_2. \tag{22}$$

By eliminating β and $\tilde{\beta}$ values, Equations (19)–(22) become

$$6\alpha_4 + 3\tilde{\alpha}_3 - 3\alpha_3 = 1, \tag{23}$$
$$6\tilde{\alpha}_4 + 3\alpha_3 - 3\tilde{\alpha}_3 = 1. \tag{24}$$

5. Symplectic PRK Methods with Effective Order 3

The flow of a Hamiltonian system (2) is symplectic and it is a well known fact that the discrete flow by symplectic Runge-Kutta methods is symplectic [14]. The PRK method M and $\tilde{M}$ for separable Hamiltonian system (2) is symplectic if the following condition is satisfied [14]

$$\text{diag}(b)\tilde{a} + a^T \text{diag}(\tilde{b}) - b\tilde{b} = 0. \tag{25}$$

Moreover, the composition of two symplectic RK methods is symplectic [10,15].

For symplectic PRK methods, the trees related to order conditions can be divided into superfluous and non-superfluous bi-color trees. Unlike RK methods, the superfluous bi-color trees of PRK methods also contribute one order condition together with one condition from non-superfluous bi-color tree [14]. Since, the underlying bi-color tree of the rooted trees t_2 and $\tilde{t}_2$ is superfluous. So, we can ignore the condition (18) because it is automatically satisfied. Moreover, the underlying bi-color trees of t_3, t_4, $\tilde{t}_3$ and $\tilde{t}_4$ are non-superfluous, we can either take α_3 or $\tilde{\alpha}_4$ and also α_4 or $\tilde{\alpha}_3$ resulting in reducing Equations (23) and (24) to $\alpha_3 = \frac{1}{3}$ and $\tilde{\alpha}_3 = \frac{1}{3}$. Now consider the following Butcher table for methods M and $\tilde{M}$ which satisfy the symplectic condition (25)

$$
\begin{array}{c|ccc}
0 & 0 & & \\
b_1 & b_1 & & \\
b_1 + b_2 & b_1 & b_2 & \\
\hline
 & b_1 & b_2 & b_3
\end{array}
\qquad
\begin{array}{c|ccc}
\tilde{b}_1 & \tilde{b}_1 & & \\
\tilde{b}_1 + \tilde{b}_2 & \tilde{b}_1 & \tilde{b}_2 & \\
\tilde{b}_1 + \tilde{b}_2 + \tilde{b}_3 & \tilde{b}_1 & \tilde{b}_2 & \tilde{b}_3 \\
\hline
 & \tilde{b}_1 & \tilde{b}_2 & \tilde{b}_3
\end{array}
\tag{26}
$$

The Equations (15)–(17) and (23)–(24) after simplification can be written in terms of elementary weights as

$$\sum_{i=1}^{3} b_i = 1, \qquad (27)$$

$$\sum_{i=1}^{3} \tilde{b}_i = 1, \qquad (28)$$

$$b_1 \tilde{b}_1 + b_2(\tilde{b}_1 + \tilde{b}_2) + b_3 = \frac{1}{2}, \qquad (29)$$

$$\tilde{b}_1^2 b_1 + b_2(\tilde{b}_1 + \tilde{b}_2)^2 + b_3(\tilde{b}_1 + \tilde{b}_2 + \tilde{b}_3)^2 = \frac{1}{3}, \qquad (30)$$

$$\tilde{b}_2 b_1^2 + \tilde{b}_3 (b_1 + b_2)^2 = \frac{1}{3}. \qquad (31)$$

To get the values of 6 unknowns from 5 equations, we have one degree of freedom. Let us take $\tilde{b}_1 = \frac{2}{3}$ and solve Equations (27)–(31) to get M and $\tilde{M}$ methods as follows

$$\left|\begin{array}{cccc} 0 & 0 & 0 \\ \frac{13+\sqrt{205}}{12} & 0 & 0 \\ \frac{13+\sqrt{205}}{12} & \frac{5}{6} & 0 \\ \hline \frac{13+\sqrt{205}}{12} & \frac{5}{6} & \frac{-11-\sqrt{205}}{12} \end{array}\right| \qquad \left|\begin{array}{ccc} \frac{2}{3} & 0 & 0 \\ \frac{2}{3} & \frac{5+\sqrt{205}}{30} & 0 \\ \frac{2}{3} & \frac{5+\sqrt{205}}{30} & \frac{5-\sqrt{205}}{30} \\ \hline \frac{2}{3} & \frac{5+\sqrt{205}}{30} & \frac{5-\sqrt{205}}{30} \end{array}\right|$$

Derivation of Starting Method

For the starting method, we have the following equations

$$\beta_1 = 0, \qquad (32)$$

$$\tilde{\beta}_1 = 0, \qquad (33)$$

$$\beta_2 = \frac{1}{2}\alpha_3 - \frac{1}{6}, \qquad (34)$$

$$\tilde{\beta}_2 = \frac{1}{2}\tilde{\alpha}_3 - \frac{1}{6}. \qquad (35)$$

The starting methods should be symplectic [16]. The solution of (32)–(35) leads us to the following symplectic staring PRK methods S and $\tilde{S}$ as

$$\left|\begin{array}{ccc} 0 & 0 & 0 \\ \frac{1}{3} & 0 & 0 \\ \frac{1}{3} & \frac{2}{5} & 0 \\ \hline \frac{1}{3} & \frac{2}{5} & -\frac{11}{15} \end{array}\right| \qquad \left|\begin{array}{ccc} \frac{1}{3} & 0 & 0 \\ \frac{1}{3} & -\frac{11}{18} & 0 \\ \frac{1}{3} & -\frac{11}{18} & \frac{5}{18} \\ \hline \frac{1}{3} & -\frac{11}{18} & \frac{5}{18} \end{array}\right|$$

6. Mutually Adjoint Symplectic Effective Order PRK Methods

A separable Hamiltonian system remains unchanged by changing the role of kinetic and potential energies, position, and momentum and also inverting the direction of time. For the two PRK method tableaux (26) being mutually adjoint, we have $b_1 = \tilde{b}_3$, $b_2 = \tilde{b}_2$ and $b_3 = \tilde{b}_1$ in Equations (27)–(31), so we have

$$\sum_{i=1}^{3} \tilde{b}_i = 1, \tag{36}$$

$$\sum_{i=1}^{3} \tilde{\tilde{b}}_i = 1, \tag{37}$$

$$\tilde{b}_3 \tilde{\tilde{b}}_1 + \tilde{b}_2(\tilde{\tilde{b}}_1 + \tilde{\tilde{b}}_2) + \tilde{\tilde{b}}_1 = \frac{1}{2}, \tag{38}$$

$$\tilde{b}_2 \tilde{\tilde{b}}_3^2 + \tilde{b}_3(\tilde{\tilde{b}}_3 + \tilde{\tilde{b}}_2)^2 = \frac{1}{3}. \tag{39}$$

Sanz-Serna suggested in [14] to take $\tilde{b}_3 = 0.91966152$, which leads us to the following main methods M and $\tilde{M}$ with effective order 3 with just 3 stages

$$\begin{array}{c|ccc}
0 & 0 & 0 & 0 \\
0.91966152 & 0 & 0 & 0 \\
0.91966152 & -0.18799162 & 0 & 0 \\
\hline
& 0.91966152 & -0.18799162 & 0.26833010
\end{array}
\quad , \quad
\begin{array}{c|ccc}
0.26833010 & 0 & 0 & 0 \\
0.26833010 & -0.18799162 & 0 & 0 \\
0.26833010 & -0.18799162 & 0.91966152 & \\
\hline
& 0.26833010 & -0.18799162 & 0.91966152
\end{array}$$

The starting methods S and $\tilde{S}$ for mutually adjoint symplectic effective order PRK method are constructed by using $B_1 = \tilde{B}_3$, $B_2 = \tilde{B}_2$, $B_3 = \tilde{B}_1$ in Equations (32) to (34) to get

$$\tilde{B}_1 + \tilde{B}_2 + \tilde{B}_3 = 0, \tag{40}$$

$$\tilde{B}_1 \tilde{B}_3 + \tilde{B}_2(\tilde{B}_1 + \tilde{B}_2) + \tilde{B}_1(\tilde{B}_1 + \tilde{B}_2 + \tilde{B}_3) \simeq 0. \tag{41}$$

By solving Equations (40) and (41) with $\tilde{B}_3 = \frac{1}{2}$, we get the following S and $\tilde{S}$ methods:

$$\begin{array}{c|ccc}
0 & 0 & 0 & 0 \\
\frac{1}{2} & 0 & 0 & 0 \\
\frac{1}{2} & -\frac{1}{4} & 0 & \\
\hline
& \frac{1}{2} & -\frac{1}{4} & \frac{1}{4}
\end{array}
\quad \text{and} \quad
\begin{array}{c|ccc}
-\frac{1}{4} & 0 & 0 \\
-\frac{1}{4} & -\frac{1}{4} & 0 \\
-\frac{1}{4} & -\frac{1}{4} & \frac{1}{2} \\
\hline
-\frac{1}{4} & -\frac{1}{4} & \frac{1}{2}
\end{array}$$

7. Numerical Experiments

The symplectic RK methods should be implicit and hence their computational cost is higher due to large number of function evaluations. On the other hand, we can use symplectic explicit PRK methods for Hamiltonian systems with lesser computational cost because of the explicit nature of the stages. Our derived effective order symplectic PRK methods are explicit and we have used MATLAB to implement them on Hamiltonian systems for the energy conservation and order confirmation of these methods.

7.1. Kepler's Two Body Problem

Consider Kepler's two body problem given as

$$u_1' = v_1, \quad u_1(0) = 1 - e,$$
$$u_2' = v_2, \quad u_2(0) = 0,$$
$$v_1' = \frac{-u_1}{r^3}, \quad v_1(0) = 0,$$
$$v_2' = \frac{-u_2}{r^3}, \quad v_2(0) = \sqrt{\frac{1+e}{1-e}}.$$

where $r = \sqrt{u_1^2 + u_2^2}$. The energy is given by,

$$H = \frac{1}{2}(v_1^2 + v_2^2) - \frac{1}{\sqrt{u_1^2 + u_2^2}}.$$

The exact solution after half revolution is

$$u_1(\pi) = 1 + e, \quad u_2(\pi) = 0, \quad v_1(\pi) = 0, \quad v_2(\pi) = \sqrt{\frac{1-e}{1+e}}.$$

To verify the effective order 3 behavior, we apply the starting methods S and $\tilde{S}$ to perturb initial values to $(\tilde{u}_1)_0$, $(\tilde{u}_2)_0$ and $(\tilde{v}_1)_0$, $(\tilde{v}_2)_0$, respectively. Then we apply the main methods M and $\tilde{M}$ for n number of iterations to $(\tilde{u}_1)_0$, $(\tilde{u}_2)_0$ and $(\tilde{v}_1)_0$, $(\tilde{v}_2)_0$, respectively and obtain the numerical solutions $(\tilde{u}_1)_n$, $(\tilde{u}_2)_n$ and $(\tilde{v}_1)_n$, $(\tilde{v}_2)_n$ computed at $t_n = t_0 + nh$ where h is the step-size. We then evaluate exact solutions at t_n to get $u_1(t_n)$, $u_2(t_n)$ and $v_1(t_n)$, $v_2(t_n)$ and perturb them using starting methods S and $\tilde{S}$ to obtain $\tilde{u}_1(t_n)$, $\tilde{u}_2(t_n)$ and $\tilde{v}_1(t_n)$, $\tilde{v}_2(t_n)$. Finally, to obtain global error, we take the difference between numerical and exact solutions, i.e., $||\tilde{u}_n - \tilde{u}(t_n)||$. Effective order 3 behavior for symplectic and mutually adjoint symplectic PRK is confirmed from Tables 5 and 6.

Table 5. Global errors and their comparison: Symplectic Effective order PRK method.

h	n	Global Error	Ratio
$\frac{\pi}{225}$	225	$7.7741637102284 \times 10^{-04}$	
			8.927465
$\frac{\pi}{450}$	450	$8.7081425338124 \times 10^{-05}$	
			8.475956
$\frac{\pi}{900}$	900	$1.02739357736254 \times 10^{-05}$	
			8.063877
$\frac{\pi}{1800}$	1800	$1.27406905909534 \times 10^{-06}$	

We calculate the ratio of global errors calculated at step length h, $\frac{h}{2}$, $\frac{h}{4}$ and $\frac{h}{8}$. For method of order p, ratio should approximately be 2^p [12].

The next experiment is to verify the energy conservation behaviour of the symplectic methods. In this experiment, the step-size $h = 2\pi/1000$ is used for 10^5 iterations. Figures 1 and 2 depict good conservation of energy with symplectic and mutually adjoint symplectic effective order PRK methods, respectively. Whereas, the energy error was bounded above by 10^{-13} and 10^{-14}, respectively.

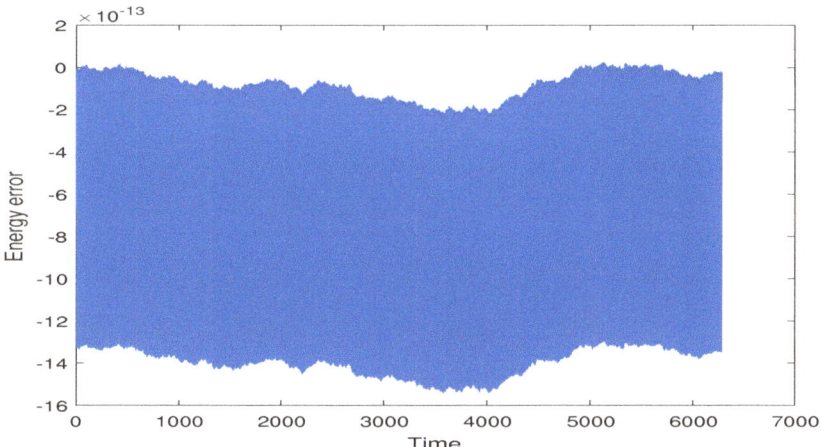

Figure 1. Energy error for the Kepler's Problem ($e = 0$) with symplectic effective PRK method using step size $h = 2\pi/1000$ for 10^5 steps.

Table 6. Global errors and their comparison: Mutually adjoint symplectic Effective order PRK method.

h	n	Global Error	Ratio
$\frac{\pi}{40}$	40	$4.635890382086 \times 10^{-03}$	
			7.963129
$\frac{\pi}{80}$	80	$5.82169473987045 \times 10^{-04}$	
			8.062262
$\frac{\pi}{160}$	160	$7.22091971134495 \times 10^{-05}$	
			7.836579
$\frac{\pi}{320}$	320	$9.21437776243649 \times 10^{-06}$	

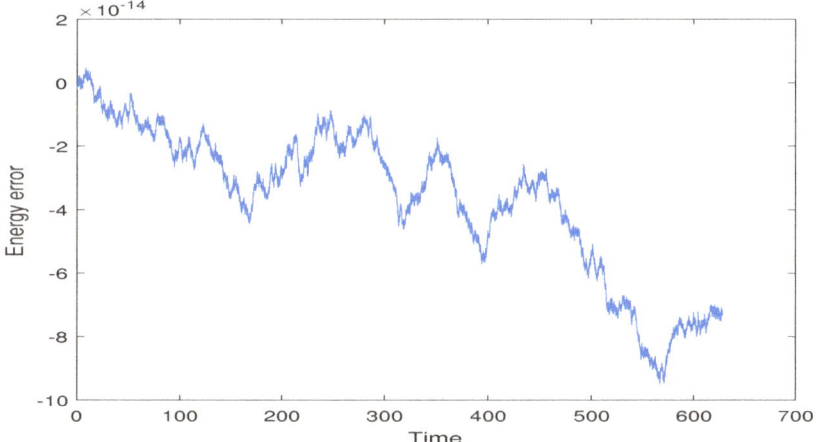

Figure 2. Energy error for the Kepler's Problem ($e = 0$) with mutually adjoint symplectic effective PRK method using step-size $h = 2\pi/1000$ for 10^5 steps.

7.2. Harmonic Oscillator

The motion of a unit mass attached to a spring with momentum u and position co-ordinates v defines the Hamiltonian system

$$v' = u \qquad u' = v$$

The energy is given by

$$H = \frac{u^2}{2} + \frac{v^2}{2}$$

The exact solution is

$$\begin{bmatrix} u(t) \\ v(t) \end{bmatrix} = \begin{bmatrix} \cos(t) & -\sin(t) \\ \sin(t) & \cos(t) \end{bmatrix} \begin{bmatrix} u(0) \\ v(0) \end{bmatrix}$$

We have applied the symplectic PRK and mutually adjoint symplectic PRK methods with step-size $h = 2\pi/1000$ with 10^5 iteration in this experiment. We have obtained good energy conservation as shown in Figures 3 and 4 by symplectic effective order PRK and mutually adjoint symplectic effective order PRK methods, respectively.

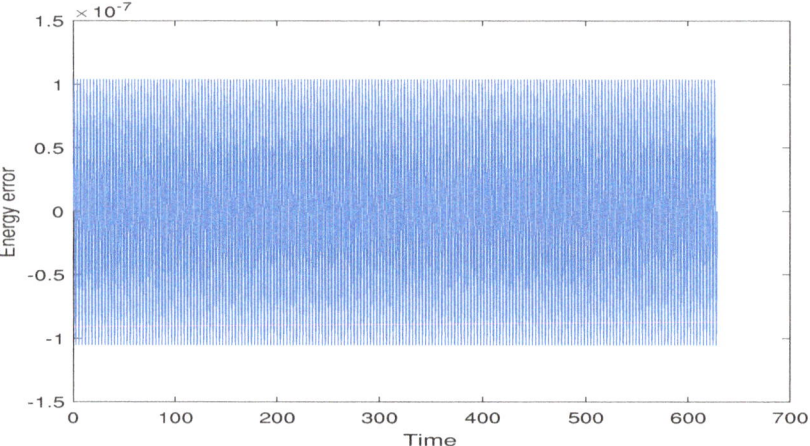

Figure 3. Energy error for the Harmonic oscillator problem ($e = 0$) with symplectic effective PRK method using step size $h = 2\pi/1000$ for 10^5 steps.

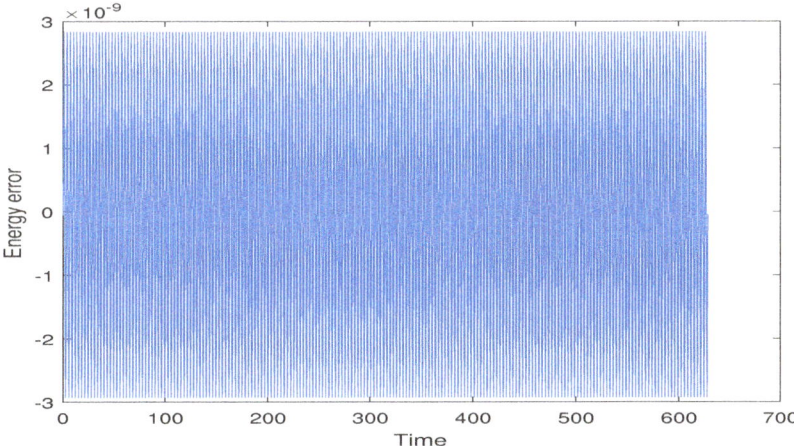

Figure 4. Energy error for the Harmonic oscillator problem ($e = 0$) with mutually adjoint symplectic effective PRK method using step size $h = 2\pi/1000$ for 10^5 steps.

8. Conclusions

In this paper, the composition of two PRK methods is elaborated in terms of Butcher tableaux as well as in terms of their elementary weight functions. The effective order conditions are provided for PRK methods and 3 stage effective order 3 symplectic PRK methods are constructed and successfully applied to separable Hamiltonian systems with good energy conservation. It is worth mentioning that we are able to construct mutually adjoint symplectic effective order 3 PRK methods with just 3 stages, whereas an equivalent method of order 3 with 4 stages is given in [16]. In dynamical solar systems, we deal with many problems which are described by Hamiltonian systems, like, Kepler's two body problem and more realistic Jovian five body problem. These symplectic methods are much useful for such types of physical phenomena to observe their positions and energy conservation.

Author Contributions: Formal analysis, J.A., A.A., S.S. and M.Y.; investigation, J.A., Y.H., S.u.R., A.A. and S.S.; software, J.A., Y.H., S.u.R and M.Y.; validation, J.A.; writing—original draft, J.A.; writing—review & editing, Y.H., S.u.R. and S.S.; conceptualization, Y.H.; methodology Y.H. and S.S.; Supervision, Y.H. and S.u.R.; visualization, Y.H., A.A. and M.Y.; resources, S.S. and M.Y.

Funding: This research received no external funding.

Conflicts of Interest: The authors declare no conflict of interest.

References

1. Sanz-Serna, J.M.; Calvo, M.P. *Numerical Hamiltonian Problems*; Chapman & Hall: London, UK, 1994.
2. Blanes, S.; Moan, P.C. Practical symplectic partitioned Runge–Kutta and Runge–Kutta–Nyström methods. *J. Comput. Appl. Math.* **2002**, *142*, 313–330. [CrossRef]
3. Chou, L.Y.; Sharp, P.W. On order 5 symplectic explicit Runge-Kutta Nyström methods. *Adv. Decis. Sci.* **2000**, *4*, 143–150. [CrossRef]
4. Geng, S. Construction of high order symplectic Runge-Kutta methods. *J. Comput. Math.* **1993**, *11*, 250–260.
5. Butcher, J.C. The effective order of Runge–Kutta methods. *Lecture Notes Math.* **1969**, *109*, 133–139.
6. Butcher, J.C. Order and effective order. *Appl. Numer. Math.* **1998**, *28*, 179–191. [CrossRef]
7. Butcher, J.C.; Chartier, P. A generalization of singly-implicit Runge-Kutta methods. *Appl. Numer. Math.* **1997**, *24*, 343–350. [CrossRef]
8. Butcher, J.C.; Chartier, P. The effective order of singly-implicit Runge-Kutta methods. *Numer. Algorithms* **1999**, *20*, 269–284. [CrossRef]

9. Butcher, J.C.; Diamantakis, M.T. DESIRE: Diagonally extended singly implicit Runge—Kutta effective order methods. *Numer. Algorithms* **1998**, *17*, 121–145. [CrossRef]
10. Butcher, J.C.; Imran, G. Symplectic effective order methods. *Numer. Algorithms* **2014**, *65*, 499–517. [CrossRef]
11. Lopez-Marcos, M.; Sanz-Serna, J.M.; Skeel, R.D. Cheap Enhancement of Symplectic Integrators. *Numer. Anal.* **1996**, *25*, 107–122.
12. Butcher, J.C. *Numerical Methods for Ordinary Differential Equations*, 2nd ed.; Wiley: Hoboken, NJ, USA, 2008.
13. Hairer, E.; Nørsett, S.P.; Wanner, G. *Solving Ordinary Differential Equations I for Nonstiff Problems*, 2nd ed.; Springer: Berlin, Germany, 1987.
14. Sanz-Serna, J.M. Symplectic integrators for Hamiltonian problems. *Acta Numer.* **1992**, *1*, 123–124. [CrossRef]
15. Habib, Y. Long-Term Behaviour of G-symplectic Methods. Ph.D. Thesis, The University of Auckland, Auckland, New Zealand, 2010.
16. Butcher, J.C.; D'Ambrosio, R. Partitioned general linear methods for separable Hamiltonian problems. *Appl. Numer. Math.* **2017**, *117*, 69–86. [CrossRef]

© 2019 by the authors. Licensee MDPI, Basel, Switzerland. This article is an open access article distributed under the terms and conditions of the Creative Commons Attribution (CC BY) license (http://creativecommons.org/licenses/by/4.0/).

Article

Explicit Integrator of Runge-Kutta Type for Direct Solution of $u^{(4)} = f(x, u, u', u'')$

Nizam Ghawadri [1], **Norazak Senu** [1,2,*], **Firas Adel Fawzi** [3], **Fudziah Ismail** [1,2] and **Zarina Bibi Ibrahim** [1,2]

[1] Institute for Mathematical Research , Universiti Putra Malaysia, Serdang 43400, Selangor, Malaysia; nizamghawadri@gmail.com (N.G.); fudziah@upm.edu.my (F.I.); zarinabb@upm.edu.my (Z.B.I.)
[2] Department of Mathematics, Universiti Putra Malaysia, Serdang 43400, Selangor, Malaysia
[3] Department of Mathematics, Faculty of Computer Science and Mathematics, University of Tikrit, Salah ad Din P.O. Box 42, Iraq; firasadil01@gmail.com
* Correspondence: norazak@upm.edu.my

Received: 30 November 2018; Accepted: 10 January 2019; Published: 16 February 2019

Abstract: The primary contribution of this work is to develop direct processes of explicit Runge-Kutta type (RKT) as solutions for any fourth-order ordinary differential equation (ODEs) of the structure $u^{(4)} = f(x, u, u', u'')$ and denoted as RKTF method. We presented the associated B-series and quad-colored tree theory with the aim of deriving the prerequisites of the said order. Depending on the order conditions, the method with algebraic order four with a three-stage and order five with a four-stage denoted as RKTF4 and RKTF5 are discussed, respectively. Numerical outcomes are offered to interpret the accuracy and efficacy of the new techniques via comparisons with various currently available RK techniques after converting the problems into a system of first-order ODE systems. Application of the new methods in real-life problems in ship dynamics is discussed.

Keywords: Runge-Kutta type methods; fourth-order ODEs; order conditions; B-series; quad-colored trees

---

1. Introduction

Fourth-order ODEs can be found in several areas of neural network engineering and applied sciences [1], fluid dynamics [2], ship dynamics [3–5], electric circuits [6] and beam theory [7,8]. In this article, we are dealing with development and explanation of the numerical process to solve fourth-order initial-value problems (IVPs) of the case:

$$u^{(4)}(x) = f(x, u(x), u'(x), u''(x)), \tag{1}$$

with initial conditions

$$u(x_0) = u_0, \quad u'(x_0) = u'_0, \quad u''(x_0) = u''_0, \quad u'''(x_0) = u'''_0, \quad x \geq x_0$$

where $u, u', u'', u''' \in \mathbb{R}^d$, $f : \mathbb{R} \times \mathbb{R}^d \times \mathbb{R}^d \times \mathbb{R}^d \to \mathbb{R}^d$ constitute continuous vector-valued functions without third derivatives. The general fourth order needs more function evaluations to be calculated, which requires extra calculation effort and extended execution time. So we have presented the explicit formulas of RKT to solve fourth-order ODEs directly of the structure $u^{(4)} = f(x, u, u', u'')$. The numerical solution is very significant to ODEs of order four that are used in various applications since the exact solutions usually do not exist. Many researchers have used classical approaches to solve higher-order ODEs through converting them to first order system of ODEs and thus using appropriate numerical approach to this arrangement (see [9–11]). However, this strategy is extremely expensive because several researchers found that converting higher-order ODEs into first-order ODE systems will increase the equation count (see [7,12,13]). Consequently, more function evaluations need to be calculated, which requires into more computational effort and longer time. Many researchers have suggested direct numerical approach to more accurate results with less calculation time (see [14–19]). Furthermore, Ibrahim et al. [20] found a process by using multi-step technique which could solve stiff differential equations of order three. Jain et al. [21] developed finite difference approach to solve ODEs of order four, all the methods discussed above are multi-step in nature. On the other hand, Mechee et al. [22,23], constructed a RK-based method for solving special third-order ODEs directly. Senu et al. [24] developed embedded explicit RKT method to directly solve special ODEs of order three. Subsequently, Hussain et al. [25] proposed RKT approach for solving the aforementioned equations, except that the latter were of order four. The main purpose of this study is using quad-colored trees theory to construct one step explicit RKT approach to solve fourth-order ODEs of the structure $u^{(4)} = f(x, u, u', u'')$ denoted as RKTF method.

The motivation of this study is to solve specific real-life problems such as ship dynamics which is special fourth-order ODE. Add to that, special method, RKTF will be considered that can solved directly special fourth-order ODEs which is more efficient than the general method because of the complexity of the method.

We organized this paper as follows: The idea of formulation of the RKTF methods to solve problem (1) is discussed in Section 2. B-series and associated quad-colored for RKTF methods are presented in Section 3. Section 4 investigates the construction of three- and four-staged RKTF methods of fourth and fifth orders, respectively. In the subsequent section, the efficiencies as well as accuracies the techniques will be compared against those of the existing methods. The ship dynamics problem is discussed in Section 6. Lastly conclusions and discussion are given in Section 7.

2. Formulation of the RKTF Methods

The s-stage Runge-Kutta type technique for IVP (1) of order four is given through the scheme as follows

$$U_i = u_n + c_i h u'_n + \frac{1}{2}c_i^2 h^2 u''_n + \frac{1}{6}c_i^3 h^3 u'''_n + h^4 \sum_{j=1}^{s} a_{ij} f(x_n + c_j h, U_j, U'_j, U''_j),$$

$$U'_i = u'_n + c_i h u''_n + \frac{1}{2}c_i^2 h^2 u'''_n + h^3 \sum_{j=1}^{s} \bar{a}_{ij} f(x_n + c_j h, U_j, U'_j, U''_j),$$

$$U''_i = u''_n + c_i h u'''_n + h^2 \sum_{j=1}^{s} \bar{\bar{a}}_{ij} f(x_n + c_j h, U_j, U'_j, U''_j),$$

$$u_{n+1} = u_n + h u'_n + \frac{1}{2}h^2 u''_n + \frac{1}{6}h^3 u'''_n + h^4 \sum_{i=1}^{s} b_i f(x_n + c_i h, U_i, U'_i, U''_i),$$

$$u'_{n+1} = u'_n + h u''_n + \frac{1}{2}h^2 u'''_n + h^3 \sum_{i=1}^{s} b'_i f(x_n + c_i h, U_i, U'_i, U''_i),$$

$$u''_{n+1} = u''_n + h u'''_n + h^2 \sum_{i=1}^{s} b''_i f(x_n + c_i h, U_i, U'_i, U''_i),$$

$$u'''_{n+1} = u'''_n + h \sum_{i=1}^{s} b'''_i f(x_n + c_i h, U_i, U'_i, U''_i). \tag{2}$$

The assumingly real new parameters $b_i, b'_i, b''_i, b'''_i, a_{ij}, \bar{a}_{ij}, \bar{\bar{a}}_{ij}$ and c_i of the RKTF method and used for $i, j = 1, 2, ..., s$. The technique is explicit if $a_{ij} = \bar{a}_{ij} = \bar{\bar{a}}_{ij} = 0$ for $i \leq j$ and it is implicit otherwise. In Kroneker's block product, the scheme is given through as follows:

$$U = e \otimes u_n + h(c \otimes u'_n) + \frac{h^2}{2}(c^2 \otimes u''_n) + \frac{h^3}{6}(c^3 \otimes u'''_n) + h^4(A \otimes I_d) F(U, U', U''),$$

$$U' = e \otimes u'_n + h(c \otimes u''_n) + \frac{h^2}{2}(c^2 \otimes u'''_n) + h^3(\bar{A} \otimes I_d) F(U, U', U''),$$

$$U'' = e \otimes u''_n + h(c \otimes u'''_n) + h^2(\bar{\bar{A}} \otimes I_d) F(U, U', U''),$$

$$u_{n+1} = u_n + h u'_n + \frac{1}{2}h^2 u''_n + \frac{1}{6}h^3 u'''_n + h^4(b^T \otimes I_d) F(U, U', U''),$$

$$u'_{n+1} = u'_n + h u''_n + \frac{1}{2}h^2 u'''_n + h^3(b'^T \otimes I_d) F(U, U', U''),$$

$$u''_{n+1} = u''_n + h u'''_n + h^2(b''^T \otimes I_d) F(U, U', U''),$$

$$u'''_{n+1} = u'''_n + h(b'''^T \otimes I_d) F(U, U', U'').$$

where, $e = [1, ..., 1]^T$, $c = [c_1, ..., c_s]^T$, $b = [b_1, ..., b_s]^T$, $b' = [b'_1, ..., b'_s]^T$, $b'' = [b''_1, ..., b''_s]^T$, $b''' = [b'''_1, ..., b'''_s]^T$, $A = [a_{ij}]^T$, $\bar{A} = [\bar{a}_{ij}]^T$, $\bar{\bar{A}} = [\bar{\bar{a}}_{ij}]^T$ denote $s \times s$ matrices while I_d denotes $d \times d$ identity matrix. The definition of all block vectors within $\mathbb{R}^{s \times d}$ are as follows:

$$U = (U_1^T, ..., U_s^T)^T,$$

$$F(U, U', U'') = \left(f(x_n + c_i h, U_i, U'_i, U''_i)^T, ..., f(x_n + c_s h, U_s, U'_s, U''_s)^T \right)^T, i = 1, 2, ..., s.$$

The RKTF methods can be presented by the Butcher tableau of scheme (2) as follows (see Table 1):

Table 1. The Butcher tableau RKTF method.

c	A	$\bar{A}$	$\bar{\bar{A}}$	
	b^T	b'^T	b''^T	b'''^T

3. B-Series and Linked Quad-Colored for RKTF Methods

This section will provide the important definitions that linked relevant theorems used in this work.

Definition 1. *The RKTF formula (2) is q-ordered if for every Equation (1) of sufficient smoothness, with respect to a proposition that $u(x_n) = u_n$, $u'(x_n) = u'_n$, $u''(x_n) = u''_n$, $u'''(x_n) = u'''_n$, the local truncation errors of the analytic solutions as well as their derivatives must fulfil the following: (see Hussain et al. [25] and Chen et al. [26])*

$$\| u(x_n + h) - u_{n+1} \| = O(h^{q+1}), \quad \| u'(x_n + h) - u'_{n+1} \| = O(h^{q+1}),$$
$$\| u''(x_n + h) - u''_{n+1} \| = O(h^{q+1}), \quad \| u'''(x_n + h) - u'''_{n+1} \| = O(h^{q+1})$$

3.1. RKTF Trees and B-Series Theory

To construct the order conditions to RKTF approach Equation (2), we are required to use autonomous formula of fourth-order IVP Equation (1)

$$u^{(4)}(x) = f(u(x), u'(x), u''(x)), \tag{3}$$

subject to initial prerequisites of

$$u(x_n) = u_n, \quad u'(x_n) = u'_n, \quad u''(x_n) = u''_n, \quad u'''(x_n) = u'''_n.$$

The IVP (1) of order four can be defined as the autonomous form through expansion of initial-value problem (1) using one-dimensioned vector $z = x$

$$z^{(4)} = 0,$$
$$u^{(4)} = f(z, u, u', u''),$$
$$z(x_n) := z_n = x_n, \ z'(x_n) := z'_n = 1, \ z''(x_n) := z''_n = 0, \ z'''(x_n) := z'''_n = 0,$$
$$u(x_n) = u_n, \ u'(x_n) = u'_n, \ u''(x_n) = u''_n, \ u'''(x_n) = u'''_n. \tag{4}$$

We will obtain the same result when the RKTF approach (2) is applied to the autonomous Equation (4) and also to the non-autonomous problem (1). Thus, we want only consider the autonomous Equation (3) (see Hussain et al. [25]). Hence, to get a common method to obtain the higher-order derivatives to the analytic solutions for Equation (3), we note that the elementary differentials up to six derivatives for $u(x)$ at $x = x_0$ are given as follow:

$$u^{(1)} = u', \ u^{(2)} = u'', \ u^{(3)} = u''', \ u^{(4)} = f, \ u^{(5)} = f'_u u' + f'_{u'} u'' + f'_{u''} u''',$$
$$u^{(6)} = f''_{uu}(u', u') + 2f''_{uu'}(u', u'') + f''_{u'u'}(u'', u'') + 2f''_{uu''}(u', u''') +$$
$$2f''_{u'u''}(u'', u''') + f''_{u''u''}(u''', u''') + f'_u u'' + f'_{u'} u''' + f'_{u''} f \tag{5}$$

Based on Hairer et al. ([9], p. 286) a better method to tackle this issue is to use graphical exemplification indicated by quad-colored trees, in addition to some amendments to the ODEs of order four. These trees contain four kinds of; "meagre" , "black ball", "white bal l", as well as "black ball inside white ball" vertices both with brackets to link them. Fairly, in these trees we use the finish "meagre vertex" to denote for all u', the finish "black-ball vertices" to denote for all u'', the finish "white ball vertex" to denote for all u''' and the finish "black-ball-within-white-ball vertex" to denote for all f, and all arc leaves of this vertex is the m-ordered f-derivative based on u, u', u''. The sign τ_1 is denoted to the first algebraic order tree, the sign τ_2 is denoted to the second algebraic order tree, the sign τ_3 is denoted to a algebraic order three tree, while τ_4 is denoted to the fourth algebraic order tree (see Figure 1).

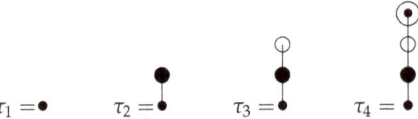

Figure 1. The quad-colored trees.

Definition 2. *The repetitively explaining for the group of quad-colored trees (RT) that gives the following: (see Hussain et al. [25] and Chen et al. [26])*

(a) *The tree τ_1 includes just one "meagre vertex" (called root) and $\tau_1 \in RT$ and also trees mentioned above τ_2, τ_3 and τ_4 are in RT.*

(b) *If $t_1, ..., t_r, t_{r+1}, ..., t_n, t_{n+1}, ..., t_m \in RT$, then $t = [t_1, ..., t_r, < t_{r+1}, ..., t_n >, < t_{n+1}, ..., t_m >]_4 \in RT$ is the tree gained through connecting $t_1, ..., t_r, t_{r+1}, ..., t_n, t_{n+1}, ..., t_m$, to "black ball inside white ball vertex" of the tree τ_4 in RT and the root of the "meagre vertex" τ_1 is at the bottom. The subscript 4 is to remind that the trees of the roots of $t_1, ..., t_r, t_{r+1}, ..., t_n, t_{n+1}, ..., t_m$ to the tree τ_4 include a series of four vertex.*

To produce the quad-colored trees we shall use these basics:

(a) The "meagre" vertex is permanently the root.
(b) A "meagre" vertex has just one kid and this kid have to be "black ball".
(c) A "black-ball" vertex has just one kid and this kid have to be "white ball".

(d) A "white ball" vertex has just one kid and this kid have to be "black ball inside white ball vertex".
(e) Each kid of a "black ball inside white ball vertex" vertex has to be "meagre".

Definition 3. *We acquaint the order $\rho(t)$ and similarity $\sigma(t)$ functions as follows: (see Hussain et al. [25])*

(a) $\rho(\tau_1) = 1, \rho(\tau_2) = 2, \rho(\tau_3) = 3, \rho(\tau_4) = 4,$
(b) $\sigma(\tau_1) = 1, \sigma(\tau_2) = 1, \sigma(\tau_3) = 1, \sigma(\tau_4) = 1,$
(c) If $t = [t_1, ..., t_r, < t_{r+1}, ..., t_n >, < t_{n+1}, ..., t_m >]_4$, $\forall t \in RT$, then $\rho(t) = 4 + \sum_{i=1}^{r} \rho(t_i) + \sum_{i=r+1}^{n}(\rho(t_i) - 1) + \sum_{i=n+1}^{m}(\rho(t_i) - 2)$ and $\sigma(t) = \prod_{i=1}^{m}(\sigma(t_i))^{\mu_i!}(\mu_1! \mu_2!...)$, where $\rho(t)$ is the number of vertices of t, $t \in RT$ and $\mu_1! \mu_2!...$ count equal trees between $t_1, ..., t_m$.

Then we can acquaint the set S_p that contain all trees RT of order p, where $\mu_i!$ is the multiplicity of t_i for $i = 1, ..., m$.

Definition 4. *The vector-valued function $F(t) : \mathbb{R}^d \times \mathbb{R}^d \times \mathbb{R}^d \times \mathbb{R}^d \to \mathbb{R}^d$ on RT is defined as the elementary differential to every tree, $t \in RT$ recursively by (see Hussain et al. [25])*

(a) $F(\emptyset)(u, u', u'', u''') = u$, $F(\tau_1)(u, u', u'', u''') = u'$, $F(\tau_2)(u, u', u'', u''') = u''$, $F(\tau_3)(u, u', u'', u''') = u'''$, $F(\tau_4)(u, u', u'', u''') = f(u, u', u'')$,
(b) $F(t) = \frac{\partial^m f}{\partial u^r \partial u''^{m-r} \partial u'''^{m-n}}\left(F(t_1)(u, u', u'', u'''), ..., F(t_m)(u, u', u'', u''')\right)$ for $t = [t_1, ..., t_r, < t_{r+1}, ..., t_n >, < t_{n+1}, ..., t_m >]_4$.

Note: we denote by $< t_{r+1}, ..., t_n >, < t_{n+1}, ..., t_m >$ the quad-colored tree whose new roots are black ball, white ball and black ball inside white ball. (see Table 2).

By the acquaint of B-series on the tri-colored trees in [27] and the acquaint of B-series on the root trees in ([28], p. 57), we expanded these theorems and definitions to RKTF formulas to grant the use qualifier of B-series on the group RT from the quad-colored trees.

Definition 5. *For a mapping $\delta : RT \cup \{\emptyset\} \to \mathbb{R}^d$, we can define format of an official series through:*

$$B(\delta, u, u', u'') = \delta(\emptyset)y + \sum_{t \in RT} \frac{h^{\rho(t)}}{\sigma(t)} \delta(t) F(t)(u, u', u'', u'''), \quad (6)$$

is named a B-series. (see Chen et al. [26]).

We will give the fundamental lemma that provides an important role in this construct as following.

Lemma 1. *Suppose δ be a function $\delta : RT \cup \{\emptyset\} \to \mathbb{R}^d$ with $\delta(\emptyset) = 1$, $\bar{\delta}$ be a function $\bar{\delta} : RT \to \mathbb{R}^d$ with $\bar{\delta}(\tau_1) = 1$ and also $\bar{\bar{\delta}}$ be a function $\bar{\bar{\delta}} : RT \to \mathbb{R}^d$ with $\bar{\bar{\delta}}(\tau_2) = 1$. Thus, $h^4 f(B(\delta, u, u', u''), B(\frac{\rho}{h}\bar{\delta}, u, u', u''), B(\frac{\rho(\rho-1)}{h^2}\bar{\bar{\delta}}, u, u', u''))$ is also B-series $h^4 f(B(\delta, u, u', u''), B(\frac{\rho}{h}\bar{\delta}, u, u', u''), B(\frac{\rho(\rho-1)}{h^2}\bar{\bar{\delta}}, u, u', u'')) = B(\delta^{(4)}, u, u', u'')$ where $\delta^{(4)}(\emptyset) = \delta^{(4)}(\tau_1) = \delta^{(4)}(\tau_2) = \delta^{(4)}(\tau_3) = \delta^{(4)}(\tau_4) = 0$, $\delta^{(4)}(\tau_4) = 1$ and for $t = [t_1, ..., t_r, < t_{r+1}, ..., t_n >, < t_{n+1}, ..., t_m >]_4$, with $\rho(t) \geq 5$, $\delta^{(4)}(t) = \prod_{i=1}^{r} \delta(t_i) \prod_{i=r+1}^{n} \rho(t_i) \bar{\delta}(t_i) \prod_{i=n+1}^{m} \rho(t_i) (\rho(t_i) - 1) \bar{\bar{\delta}}(t_i)$.*

Table 2. Quad-colored trees of orders up to six, elementary differentials and associated functions.

Order $\rho(t)$	t	Tree	α(t)	Density	Elementary weight	Elementary differential $F(t)(u,u',u'',u''')$
0	φ	φ	1	1		u
1	τ_1		1	1		u'
2	τ_2		1	2		u''
3	τ_3		1	6		u'''
4	τ_4		1	24	e	f
5	t_{51}		1	120	c	$f'_u \, u'$
5	t_{52}		1	120	$\frac{1}{2} c^2$	$f'_{u'} \, u''$
5	t_{53}		1	120	$\frac{1}{6} c^3$	$f'_{u''} \, u'''$
6	t_{61}		1	360	c^2	$f''_{uu} \, (u',u')$
6	t_{62}		2	360	$\frac{1}{2} c^3$	$f''_{uu'} \, (u',u'')$

Table 2. Cont.

Order $\rho(t)$	t	Tree	α(t)	Density $\Upsilon(t)$	Elementary weight $\Phi(t)$	Elementary differential $F(t)(u,u',u'',u''')$
6	t_{63}		1	360	$\frac{1}{4} c^4$	$f''_{u'u'}(u'',u'')$
6	t_{64}		2	360	$\frac{1}{6} c^4$	$f''_{uu''}(u',u''')$
6	t_{65}		2	360	$\frac{1}{12} c^5$	$f''_{u'u''}(u'',u''')$
6	t_{66}		1	360	$\frac{1}{36} c^6$	$f''_{u''u''}(u''',u''')$
6	t_{67}		1	720	$\frac{1}{2} c^2$	$f'_u u''$
6	t_{68}		1	720	$\frac{1}{6} c^3$	$f'_{u'} u'''$
6	t_{69}		1	720	A	$f'_{u''} f$

Note: In this table, density is denoted as $\gamma(t)$ and elementary weight is denoted as $\eta(t)$.

Proof. By assumption, $B(\delta, u, u', u'') = u + O(h)$, $B(\frac{\rho}{h}\bar{\delta}, y, u', u'') = u' + O(h)$ and $B(\frac{\rho(\rho-1)}{h^2}\bar{\bar{\delta}}, u, u', u'') = u'' + O(h)$. Thus, the Taylor expansion of $f(B(\delta, u, u', u''), B(\frac{\rho}{h}\bar{\delta}, u, u', u''), B(\frac{\rho(\rho-1)}{h^2}\bar{\bar{\delta}}, u, u', u''))$ shows that $f(B(\delta, u, u', u''), B(\frac{\rho}{h}\bar{\delta}, u, u', u''), B(\frac{\rho(\rho-1)}{h^2}\bar{\bar{\delta}}, u, u', u'')) = h^4 f(u, u', u'') + O(h^5)$ which implies that $\delta^{(4)}(\emptyset) = \delta^{(4)}(\tau_1) = \delta^{(4)}(\tau_2) = \delta^{(4)}(\tau_3) = 0, \delta^{(4)}(\tau_4) = 1$.

Depend on the proof in Hairer et al. [28], we have

$$h^4 f(B(\delta, u, u', u''), B(\frac{\rho}{h}\bar{\delta}, y, u', u''), B(\frac{\rho(\rho-1)}{h^2}\bar{\bar{\delta}}, u, u', u'')) =$$

$$h^4 \sum_{m \geq 0} \cdots \sum_{n=0}^{m} \frac{m!}{r!(n-r)!(m-n)!} \frac{\partial^m f}{\partial u^r \partial u'^{n-r} \partial u''^{m-n}}(u, u', u'')(B(\delta, u, u'', u'') - u)^r$$

$$(B(\frac{\rho}{h}\bar{\delta}, u, u'', u'') - u')^{n-r} (B(\frac{\rho(\rho-1)}{h^2}\bar{\bar{\delta}}, u, u'', u'') - u'')^{m-n}$$

$$= h^4 \sum_{m \geq 0} \cdots \sum_{n=0}^{m} \frac{m!}{r!(n-r)!(m-n)!} \sum_{t_1 \in RT} \cdots \sum_{t_r \in RT} \sum_{t_{r+1} \in RT \setminus \{\tau_1, \tau_2, \tau_3\}} \cdots \sum_{t_n \in RT \setminus \{\tau_1, \tau_2, \tau_3\}}$$

$$\sum_{t_{n+1} \in RT \setminus \{\tau_1, \tau_2, \tau_3\}} \cdots \sum_{t_m \in RT \setminus \{\tau_1, \tau_2, \tau_3\}} \frac{h^{\rho(t_1)+\rho(t_2)+\cdots\rho(t_m)-(m-n)}}{\sigma(t_1)\cdots\sigma(t_m)} \cdot \rho(t_n+1)\cdots\rho(t_m) \delta(t_1)\cdots$$

$$\delta(t_r) \bar{\delta}(t_r+1)\cdots\bar{\delta}(t_n) \bar{\bar{\delta}}(t_n+1)\cdots\bar{\bar{\delta}}(t_m) \frac{\partial^m f}{\partial u^r \partial u'^{n-r} \partial u''^{m-n}}(u, u', u'')(F(t_1)(u, u', u'', u'''))$$

$$\cdots F(t_m)(u, u', u'', u''')) = \sum_{m \geq 0} \sum_{r=0}^{m} \sum_{t_1 \in RT} \cdots \sum_{t_r \in RT} \sum_{t_{r+1} \in RT \setminus \{\tau_1, \tau_2, \tau_3\}} \cdots$$

$$\sum_{t_n \in RT \setminus \{\tau_1, \tau_2, \tau_3\}} \sum_{t_{n+1} \in RT \setminus \{\tau_1, \tau_2, \tau_3\}} \cdots \sum_{t_m \in RT \setminus \{\tau_1, \tau_2, \tau_3\}} \frac{h^{\rho(t)}}{\sigma(t)} \cdot \frac{m! \mu_1! \mu_2! \cdots \xi_1! \xi_2! \cdots \zeta_1! \zeta_2! \cdots}{r!(n-r)!(m-n)!}$$

$$\rho(t_n+1)\cdots\rho(t_m) \delta(t_1)\cdots\delta(t_r) \bar{\delta}(t_r+1)\cdots\bar{\delta}(t_n) \bar{\bar{\delta}}(t_n+1)\cdots\bar{\bar{\delta}}(t_m) F(t)(u, u', u'', u''')$$

$$= \sum_{t \in RT, \rho(t) \geq 5} \frac{h^{\rho(t)}}{\sigma(t)} \rho(t_n+1)\cdots\rho(t_m) \delta(t_1)\cdots\delta(t_r) \bar{\delta}(t_r+1)\cdots\bar{\delta}(t_n) \bar{\bar{\delta}}(t_n+1)\cdots$$

$$\bar{\bar{\delta}}(t_m) F(t)(u, u', u'', u''')$$

where, one equality $t = [t_1, ..., t_r, < t_{r+1}, ..., t_n >, < t_{n+1}, ..., t_m >]_4$, and the number of methods of ordering the subtrees $t_1, ..., t_m$ in $t = [t_1, ..., t_r, < t_{r+1}, ..., t_n >, < t_{n+1}, ..., t_m >]_4$, i.e., the multiplicity of $t = [t_1, ..., t_r, < t_{r+1}, ..., t_n >, < t_{n+1}, ..., t_m >]_4$ is $\frac{r!(q-r)!(n-q)!(m-n)!}{m! \mu_1! \mu_2! \cdots \xi_1! \xi_2! \cdots \zeta_1! \zeta_2! \cdots}$, $\mu_1, \mu_2, \ldots$ count equal trees between $t_1, ..., t_r$, $\xi_1, \xi_2, \ldots$ count equal trees between $t_{r+1}, t_{r+2}, ..., t_n$ and $\zeta_1, \zeta_2, \ldots$ count equal trees between $t_{n+1}, t_{r+2}, ..., t_m$ we get

$$\delta^{(4)} = \prod_{i=1}^{r} \delta(t_i) \prod_{i=r+1}^{n} \rho(t_i) \bar{\delta}(t_i) \prod_{i=n+1}^{m} \rho(t_i) (\rho(t_i)-1) \bar{\bar{\delta}}(t_i).$$

Then we have

$$h^4 f(B(\delta, u, u', u''), B(\frac{\rho}{h}\bar{\delta}, y, u', u''), B(\frac{\rho(\rho-1)}{h^2}\bar{\bar{\delta}}, u, u', u'')) =$$

$$\sum_{t \in RT} \frac{h^{\rho(t)}}{\sigma(t)} \delta^{(4)} F(t)(u, u', u'', u''') = B(\delta^{(4)}, u, u', u'').$$

□

Theorem 1. *Suppose that the analytic solution $u(x_0 + h)$ of the form (3) is B-series, $B(e, u_0, u'_0, u''_0)$ with a real function e defined on $RT \cup \{\varnothing\}$. Then*

$$e(\varnothing) = 1, \quad e(\tau_1) = 1, \quad e(\tau_2) = \frac{1}{2}, \quad e(\tau_3) = \frac{1}{6}, \quad e(\tau_4) = \frac{1}{24},$$

and for $t = [t_1, ..., t_r, < t_{r+1}, ..., t_n >, < t_{n+1}, ..., t_m >]_4$,

$$e(t) = \frac{1}{\rho(t)(\rho(t)-1)(\rho(t)-2)(\rho(t)-3)} \prod_{i=1}^{r} e(t_i) \prod_{i=r+1}^{n} \rho(t_i) e(t_i)$$

$$\prod_{i=n+1}^{m} \rho(t_i)(\rho(t_i)-1)e(t_i).$$

Proof.

$$u(x_0 + h) = B(e, u_0, u_0', u_0'')$$
$$= e(\emptyset)u_0 + he(\tau_1)u_0' + h^2 e(\tau_2)u_0'' + h^3 e(\tau_3)u_0''' + h^4 e(\tau_4) f(u_0, u_0', u_0'')$$
$$+ \sum_{t_{r+1} \in RT \setminus \{\tau_1, \tau_2, \tau_3, \tau_4\}} \frac{h^{\rho(t)}}{\sigma(t)} e(t) F(u_0, u_0', u_0'', u_0'''),$$

Thus, the first fourth derivative of $u(x_0 + h)$ is presented by

$$(u(x_0+h))' = \frac{d}{dh}[u(x_0+h)] = e(\tau_1)u_0' + 2he(\tau_2)u_0'' + 3h^2 e(\tau_3)u_0''' + 4h^3 e(\tau_4) f(u_0, u_0', u_0'')$$
$$+ \sum_{t_{r+1} \in RT \setminus \{\tau_1, \tau_2, \tau_3, \tau_4\}} \frac{\rho(t) h^{\rho(t)-1}}{\sigma(t)} e(t) F(u_0, u_0', u_0'', u_0''') = B(\frac{\rho e}{h}, u_0, u_0', u_0''), \quad (7)$$

$$(u(x_0+h))^{(2)} = \frac{d^2}{dh^2}[u(x_0+h)] = 2e(\tau_2)u_0'' + 6he(\tau_3)u_0''' + 12h^2 e(\tau_4) f(u_0, u_0', u_0'')$$
$$+ \sum_{t_{r+1} \in RT \setminus \{\tau_1, \tau_2, \tau_3, \tau_4\}} \frac{\rho(t)(\rho(t)-1)h^{\rho(t)-2}}{\sigma(t)} e(t) F(u_0, u_0', u_0'', u_0'''))$$
$$= B(\frac{\rho(\rho-1)e}{h^2}, u_0, u_0', u_0''), \quad (8)$$

$$(u(x_0+h))^{(3)} = \frac{d^3}{dh^3}[u(x_0+h)] = 6e(\tau_3)u_0''' + 24he(\tau_4) f(u_0, u_0', u_0'')$$
$$+ \sum_{t_{r+1} \in RT \setminus \{\tau_1, \tau_2, \tau_3, \tau_4\}} \frac{\rho(t)(\rho(t)-1)(\rho(t)-2)h^{\rho(t)-3}}{\sigma(t)} e(t) F(u_0, u_0', u_0'', u_0''')$$
$$= B(\frac{\rho(\rho-1)(\rho-2)e}{h^3}, u_0, u_0', u_0'''),$$

$$(u(x_0+h))^{(4)} = \frac{d^4}{dh^4}[u(x_0+h)] = 24e(\tau_4) f(u_0, u_0', u_0'')$$
$$+ \sum_{t_{r+1} \in RT \setminus \{\tau_1, \tau_2, \tau_3, \tau_4\}} \frac{\rho(t)(\rho(t)-1)(\rho(t)-2)(\rho(t)-3)h^{\rho(t)-4}}{\sigma(t)} e(t) F(u_0, u_0', u_0'', u_0''')$$
$$= B(\frac{\rho(\rho-1)(\rho-2)(\rho-3)e}{h^4}, u_0, u_0', u_0''), \quad (9)$$

Moreover, of Lemma 1, we have

$$f(B(e, u, u', u''), B(\frac{\rho}{h} e, u, u', u''), B(\frac{\rho(\rho-1)}{h^2} e, u, u', u'')) = e^{(4)}(\tau_4) f(u_0, u_0', u_0'')$$
$$+ \sum_{t \in RT \setminus \{\tau_1, \tau_2, \tau_3, \tau_4\}} \frac{h^{\rho(t)-4}}{\sigma(t)} e^{(4)}(t) F(u_0, u_0', u_0'', u_0'''), \quad (10)$$

where $e^{(4)}(\tau_4) = 1$ and $t = [t_1, ..., t_r, < t_{r+1}, ..., t_n >, < t_{n+1}, ..., t_m >]_4 \in RT \setminus \{\tau_1, \tau_2, \tau_3, \tau_4\}$,

$$e^{(4)}(t) = \prod_{i=1}^{r} e(t_i) \prod_{i=r+1}^{n} \rho(t_i) e(t_i) \prod_{i=n+1}^{m} \rho(t_i)(\rho(t_i) - 1) e(t_i),$$

Inserting (9) and (10) to Equation (3), then depending on the both sides, we compare the coefficients of the same elementary differential to obtain

$$e(\tau_4) = \frac{1}{24},$$

and $t = [t_1, ..., t_r, < t_{r+1}, ..., t_n >, < t_{n+1}, ..., t_m >]_4 \in RT \setminus \{\tau_1, \tau_2, \tau_3, \tau_4\}$,

$$e(t) = \frac{1}{\rho(t)(\rho(t) - 1)(\rho(t) - 2)(\rho(t) - 3)} \prod_{i=1}^{r} e(t_i) \prod_{i=r+1}^{n} \rho(t_i) e(t_i) \prod_{i=n+1}^{m} \rho(t_i)(\rho(t_i) - 1) e(t_i).$$

lastly, depending on the Taylor series expansions of $u(x_0 + h)$ about $h = 0, e(\emptyset) = e(\tau_1) = 1$, $e(\tau_2) = \frac{1}{2}, e(\tau_3) = \frac{1}{6}, e(\tau_4) = \frac{1}{24}$. □

$\forall t \in RT$, we lead to write the density as follows $\gamma(t) = \frac{1}{e(t)}$ and also write non-negative integer as follows $\alpha(t) = \frac{\rho(t)!}{\sigma(t)\gamma(t)}$. Thus, from Theorem 1 we have two propositions that we will mention below.

Proposition 1. $\forall t \in RT$, the density $\gamma(t)$ is the non-negative integer valued function on RT satisfying. (see Hussain et al. [25] and Chen et al. [26])

(i) $\gamma(\tau_1) = 1$, $\gamma(\tau_2) = 2$, $\gamma(\tau_3) = 6$, $\gamma(\tau_4) = 24$,
(ii) $t = [t_1, ..., t_r, < t_{r+1}, ..., t_n >, < t_{n+1}, ..., t_m >]_4 \in RT$,

$$\gamma(t) = \rho(t) (\rho(t) - 1)(\rho(t) - 2)(\rho(t) - 3) \prod_{i=1}^{r} \gamma(t_i) \prod_{i=r+1}^{n} \frac{\gamma(t_i)}{\rho(t_i)} \prod_{i=n+1}^{m} \frac{\gamma(t_i)}{\rho(t_i)(\rho(t_i) - 1)},$$

Proposition 2. $\forall t \in RT$, $\alpha(t)$ is the positive-integer satisfying. (see Chen et al. [26])

(i) $\alpha(t_1) = 1$, $\alpha(t_2) = 1$, $\alpha(t_3) = 1$, $\alpha(t_4) = 1$,
(ii) $t = [t_1^{\mu_1}, ..., t_r^{\mu_r}, < t_{r+1}^{\mu_{r+1}}, ..., t_n^{\mu_n} >, , < t_{n+1}^{\mu_{n+1}}, ..., t_m^{\mu_m} >]_4 \in RT$, with $t_1, ..., t_r$ distinct and $t_{r+1}, ..., t_n$ distinct, $t_{n+1}, ..., t_m$ distinct,

$$\alpha(t) = (\rho(t) - 4)! \prod_{i=1}^{r} \frac{1}{\mu_i!} \left(\frac{\alpha(t_i)}{\rho(t_i)!}\right)^{\mu_i} \prod_{i=r+1}^{n} \frac{1}{\mu_i!} \left(\frac{\alpha(t_i)}{(\rho(t_i) - 1)!}\right)^{\mu_i} \prod_{i=n+1}^{m} \frac{1}{\mu_i!} \left(\frac{\alpha(t_i)}{(\rho(t_i) - 2)!}\right)^{\mu_i},$$

where μ_i is the multiplicity of $t_i, i = 1, ..., m$.

Then the B-series (6) can be written as follows:

$$B(\delta, u, u', u'') = \delta(\emptyset) y + \sum_{t \in RT} \frac{h^{\rho(t)}}{\rho(t)!} \delta(t) \gamma(t) \alpha(t) F(t)(u, u', u'', u'''), \quad (11)$$

and $f(B(\delta, u, u', u''), B(\frac{\rho}{h}\bar{\delta}, y, u', u''), B(\frac{\rho(\rho-1)}{h^2}\bar{\bar{\delta}}, u, u', u''))$, can be expressed as

$$f(B(\delta, u, u', u''), B(\frac{\rho}{h}\bar{\delta}, y, u', u''), B(\frac{\rho(\rho-1)}{h^2}\bar{\bar{\delta}}, u, u', u'')) = \sum_{t \in RT \setminus \{\tau_1, \tau_2, \tau_3\}} \frac{h^{\rho(t)-4}}{\rho(t)!} \delta^{(4)} \gamma(t) \alpha(t) F(t)(u, u', u'', u''').\quad(12)$$

3.2. B-Series of the Exact Solution and Exact Derivative

Depending on the former analysis, we can present the theorem as following

Theorem 2. *The analytic solution $u(x_0 + h)$ of the problem (3) and the derivative $u'(x_0 + h)$, $u''(x_0 + h)$, $u'''(x_0 + h)$ have B-series respectively as follows,*

$$u(x_0 + h) = u_0 + \sum_{t \in RT} \frac{h^{\rho(t)}}{\rho(t)!} \alpha(t) F(t)(u_0, u'_0, u''_0, u'''_0) = B\left(\frac{\alpha(t)\sigma}{\rho!}, u_0, u'_0, u''_0\right)$$
$$= B\left(\frac{1}{\gamma(t)}, u_0, u'_0, u''_0\right),\quad(13)$$

$$u'(x_0 + h) = \sum_{t \in RT} \frac{h^{\rho(t)-1}}{(\rho(t)-1)!} \alpha(t) F(t)(u_0, u'_0, u''_0, u'''_0) = B\left(\frac{\alpha(t)\sigma}{h(\rho-1)!}, u_0, u'_0, u''_0\right)$$
$$= B\left(\frac{\rho}{h\gamma(t)}, u_0, u'_0, u''_0\right),\quad(14)$$

$$u''(x_0 + h) = \sum_{t \in RT} \frac{h^{\rho(t)-2}}{(\rho(t)-2)!} \alpha(t) F(t)(u_0, u'_0, u''_0, u'''_0) = B\left(\frac{\alpha(t)\sigma}{h^2(\rho-2)!}, u_0, u'_0, u''_0\right)$$
$$= B\left(\frac{\rho(\rho-1)}{h^2\gamma(t)}, u_0, u'_0, u''_0\right),\quad(15)$$

$$u'''(x_0 + h) = \sum_{t \in RT} \frac{h^{\rho(t)-3}}{(\rho(t)-3)!} \alpha(t) F(t)(u_0, u'_0, u''_0, u'''_0) = B\left(\frac{\alpha(t)\sigma}{h^3(\rho-3)!}, u_0, u'_0, u''_0\right)$$
$$= B\left(\frac{\rho(\rho-1)(\rho-2)}{h^3\gamma(t)}, u_0, u'_0, u''_0\right).\quad(16)$$

The proof is given by Hussain et al. [25]

3.3. B-Series of the Numerical Solution and Numerical Derivative

So as to constitute the B-series for the numerical solution u_1 and the numerical derivative u'_1, u''_1, u'''_1 of the form (3) created by the RKTF approach (2), we suppose that U_i, U'_i and U''_i in Equation (2) can be developed as B-series $U_i = B(\psi_i, u_0, u'_0, u''_0)$, $U'_i = B(\frac{\rho}{h}\bar{\psi}_i, u_0, u'_0, u''_0)$ and $U''_i = B(\frac{\rho(\rho-1)}{h^2}\bar{\bar{\psi}}_i, u_0, u'_0, u''_0)$ respectively. Then the first-three equations in the scheme (2) are as follows,

$$B(\psi_i, u_0, u'_0, u''_0) = u_0 + c_i h u'_0 + \frac{1}{2} c_i^2 h^2 u''_0 + \frac{1}{6} c_i^3 h^3 u'''_0$$
$$+ h^4 \sum_{j=1}^{s} a_{ij} f(B(\psi_i, u_0, u'_0, u''_0), B(\frac{\rho}{h}\bar{\psi}_i, u_0, u'_0, u''_0), B(\frac{\rho(\rho-1)}{h^2}\bar{\bar{\psi}}_i, u_0, u'_0, u''_0)),$$

$$B\left(\frac{\rho}{h}\bar{\psi}_i, u_0, u'_0, u''_0\right) = u'_0 + c_i h u''_0 + \frac{1}{2}c_i^2 h^2 u'''_0$$
$$+ h^3 \sum_{j=1}^{s} \bar{a}_{ij} f\left(B(\psi_i, u_0, u'_0, u''_0), B\left(\frac{\rho}{h}\bar{\psi}_i, u_0, u'_0, u''_0\right), B\left(\frac{\rho(\rho-1)}{h^2}\bar{\bar{\psi}}_i, u_0, u'_0, u''_0\right)\right),$$

$$B\left(\frac{\rho(\rho-1)}{h^2}\bar{\bar{\psi}}_i, u_0, u'_0, u''_0\right) = u''_0 + c_i h u'''_0$$
$$+ h^2 \sum_{j=1}^{s} \bar{\bar{a}}_{ij} f\left(B(\psi_i, u_0, u'_0, u''_0), B\left(\frac{\rho}{h}\bar{\psi}_i, u_0, u'_0, u''_0\right), B\left(\frac{\rho(\rho-1)}{h^2}\bar{\bar{\psi}}_i, u_0, u'_0, u''_0\right)\right),$$

by (11) and (12) the former two equations can be presented as

$$\psi_i(\emptyset) u_0 + \sum_{t \in RT} \frac{h^{\rho(t)}}{\rho(t)!} \psi_i(t) \gamma(t) \alpha(t) F(t)(u, u', u'', u''') = u_0 + c_i h u'_0 + \frac{1}{2}c_i^2 h^2 u''_0$$
$$+ \frac{1}{6}c_i^3 h^3 u'''_0 + h^4 \sum_{j=1}^{s} \sum_{t \in RT \setminus \{\tau_1, \tau_2, \tau_3\}} \frac{h^{\rho(t)}}{\rho(t)!} a_{ij} \psi_j^{(4)} \gamma(t) \alpha(t) F(t)(u_0, u'_0, u''_0, u'''_0),$$

$$\sum_{t \in RT} \frac{h^{\rho(t)-1}}{(\rho(t)-1)!} \bar{\psi}_i(t) \gamma(t) \alpha(t) F(t)(u_0, u'_0, u''_0, u'''_0) = u'_0 + c_i h u''_0 + \frac{1}{2}c_i^2 h^2 u'''_0 + h^3$$
$$\sum_{j=1}^{s} \sum_{t \in RT \setminus \{\tau_1, \tau_2, \tau_3\}} \frac{h^{\rho(t)-1}}{\rho(t)!} \bar{a}_{ij} \psi_j^{(4)} \gamma(t) \alpha(t) F(t)(u_0, u'_0, u''_0, u'''_0),$$

$$\sum_{t \in RT} \frac{h^{\rho(t)-2}}{(\rho(t)-2)!} \bar{\bar{\psi}}_i(t) \gamma(t) \alpha(t) F(t)(u_0, u'_0, u''_0, u'''_0) = u''_0 + c_i h u'''_0$$
$$h^2 \sum_{j=1}^{s} \sum_{t \in RT \setminus \{\tau_1, \tau_2, \tau_3\}} \frac{h^{\rho(t)-2}}{\rho(t)!} \bar{\bar{a}}_{ij} \psi_j^{(4)} \gamma(t) \alpha(t) F(t)(u_0, u'_0, u''_0, u'''_0).$$

It follows that

$$\psi_i(\emptyset) = 1, \quad \psi_i(\tau_1) = c_i, \quad \psi_i(\tau_2) = \frac{1}{2}c_i^2, \quad \psi_i(\tau_3) = \frac{1}{6}c_i^3,$$
$$\psi_i(\tau_4) = \sum a_{ij} \psi_j^{(4)}(\tau_4) = \sum_{j=1}^{s} a_{ij}, \qquad (17)$$

$$\bar{\psi}_i(\tau_1) = 1, \quad \bar{\psi}_i(\tau_2) = c_i, \quad \bar{\psi}_i(\tau_3) = \frac{1}{2}c_i^2, \quad \bar{\psi}_i(\tau_4) = \frac{1}{4}\sum_{j=1}^{s} \bar{a}_{ij},$$
$$\bar{\bar{\psi}}_i(\tau_2) = 1, \quad \bar{\bar{\psi}}_i(\tau_3) = c_i, \quad \bar{\bar{\psi}}_i(\tau_4) = \frac{1}{12}\sum \bar{\bar{a}}_{ij}, \qquad (18)$$

and

$$\psi_i(t) = \sum_{j=1}^{s} a_{ij} \psi_j^{(4)}(t), \quad \bar{\psi}_i(t) = \sum_{j=1}^{s} \frac{\bar{a}_{ij}}{\rho(t)} \psi_j^{(4)}(t), \quad \bar{\bar{\psi}}_i(t) = \sum_{j=1}^{s} \frac{\bar{\bar{a}}_{ij}}{\rho(t)(\rho(t)-1)} \psi_j^{(4)}(t), \qquad (19)$$

furthermore, for trees $t = [t_1, ..., t_r, < t_{r+1}, ..., t_n >, < t_{n+1}, ..., t_m >]_4 \in RT$ and $\rho(t) \geq 5$, Lemma 5 gives

$$\psi_j^{(4)}(t) = \prod_{i=1}^{r} \psi_j(t_i) \prod_{i=r+1}^{n} \rho(t_i) \bar{\psi}_j \prod_{i=n+1}^{m} \rho(t_i)(\rho(t_i)-1) \bar{\bar{\psi}}_j, \qquad (20)$$

inserting (19) into (20) we obtain:

$$\psi_j^{(4)}(t) = \prod_{i=1}^{r}\left[\sum_{k=1}^{s} a_{jk}\psi_k^{(4)}(t_i)\right] \prod_{i=r+1}^{n}\left[\sum_{k=1}^{s} \bar{a}_{jk}\bar{\psi}_k^{(4)}(t_i)\right] \prod_{i=n+1}^{m}\left[\sum_{k=1}^{s} \bar{\bar{a}}_{jk}\bar{\bar{\psi}}_k^{(4)}(t_i)\right]. \quad (21)$$

We denote $\psi_j^{(4)}(t_i) = \eta_j(t)$, for all trees $t = [t_1, ..., t_r, < t_{r+1}, ..., t_n >, < t_{n+1}, ..., t_m >]_4 \in RT$ and $\rho(t) \geq 5$.

Thus, (21) can be written as follows,

$$\eta_j(t) = \prod_{i=1}^{r}\left[\sum_{k=1}^{s} a_{jk}\eta_k(t_i)\right] \prod_{i=r+1}^{n}\left[\sum_{k=1}^{s} \bar{a}_{jk}\bar{\eta}_k(t_i)\right] \prod_{i=n+1}^{m}\left[\sum_{k=1}^{s} \bar{\bar{a}}_{jk}\bar{\bar{\eta}}_k(t_i)\right].$$

Commonly, the next significant lemma yields the values of $\eta_j(\tau)$ for each tree belonging to $RT\setminus\{\tau_1, \tau_2, \tau_3\}$

Lemma 2. *We can compute the function $\eta_j(t)$ on $\in RT\setminus\{\tau_1, \tau_2, \tau_3\}$ recursively.*

(i) $\eta_j(\tau_4) = 1$
(ii) for $t = [\tau_1^{\mu_1}, ..., t_r^{\mu_r}, < t_{r+1}^{\mu_{r+1}}, ..., t_n^{\mu_n} >, < t_{n+1}^{\mu_{n+1}}, ..., t_m^{\mu_m} >]_4 \in RT$ with $t_4, ..., t_r$ distinct and different from τ_1, τ_2, τ_3, and $, t_{r+1}, ..., t_n$ distinct, $t_{n+1}, ..., t_m$ distinct,

$$\eta_i(t) = \frac{1}{2^{\mu_2} 6^{\mu_3}} c_i^{\mu_1 + 2\mu_2 + 3\mu_3} \prod_{k=4}^{r}\left[\sum_{j=1}^{s} a_{ij}\eta_j(t_k)\right]^{\mu_k} \cdot \prod_{k=r+1}^{n}\left[\sum_{j=1}^{s} \bar{a}_{ij}\eta_j(t_k)\right]^{\mu_k}$$
$$\prod_{k=n+1}^{m}\left[\sum_{j=1}^{s} \bar{\bar{a}}_{ij}\eta_j(t_k)\right]^{\mu_k},$$

where, μ_1, μ_2, μ_3 is the multiplicity of τ_1, τ_2, τ_3 respectively and μ_k is the multiplicity of t_k for $k = 4, .., n$.

Here, we define the vector $\eta(t) = (\eta_1(t), ..., \eta_s(t))^T$ for $t \in RT\setminus\{\tau_1, \tau_2, \tau_3\}$

(i) The initial weight linked to u_{n+1} is denoted by $\varphi(t) = \sum b_i \eta_i(t) = b^T \eta(t)$
(ii) $\varphi'(t)$ is denoted to the initial weight linked with u'_{n+1} and written as follows:
$\varphi'(t) = \sum_{i=1}^{s} b'_i \eta_i(t) = b'^T \eta(t)$
(iii) $\varphi''(t)$ is denoted to the initial weight linked with u''_{n+1} and written as follows:
$\varphi''(t) = \sum_{i=1}^{s} b''_i \eta_i(t) = b''^T \eta(t)$.
(iv) $\varphi'''(t)$ is denoted to the initial weight linked with u'''_{n+1} and written as follows:
$\varphi'''(t) = \sum_{i=1}^{s} b'''_i \eta_i(t) = b'''^T \eta(t)$.

Theorem 3. *The numerical solution u_1 and the numerical derivative u'_1, u''_1, u'''_1 of Equation (3) produced by the RKTF approach (2) have the following B-series*

$$u_1(x_0 + h) = u_0 + h u'_0 + \frac{1}{2}h^2 u''_0 + \frac{1}{6}h^3 u'''_0 + \sum_{t \in RT\setminus\{\tau_1, \tau_2, \tau_3\}} \frac{h^{\rho(t)}}{\rho(t)!} \varphi(t)\, \gamma(t)\, \alpha(t)\, F(t)(u_0, u'_0, u''_0, u'''_0),$$

$$u_1'(x_0+h) = u_0' + h u_0'' + \frac{1}{2}h^2 u_0''' + \sum_{t \in RT \setminus \{\tau_1,\tau_2,\tau_3\}} \frac{h^{\rho(t)-1}}{\rho(t)!} \varphi'(t)\, \gamma(t)\, \alpha(t)\, F(t)(u_0, u_0', u_0'', u_0'''),$$

$$u_1''(x_0+h) = u_0'' + h u_0''' + \sum_{t \in RT \setminus \{\tau_1,\tau_2,\tau_3\}} \frac{h^{\rho(t)-2}}{\rho(t)!} \varphi''(t)\, \gamma(t)\, \alpha(t)\, F(t)(u_0, u_0', u_0'', u_0'''),$$

$$u_1'''(x_0+h) = u_0''' + \sum_{t \in RT \setminus \{\tau_1,\tau_2,\tau_3\}} \frac{h^{\rho(t)-3}}{\rho(t)!} \varphi'''(t)\, \gamma(t)\, \alpha(t)\, F(t)(u_0, u_0', u_0'', u_0''').$$

Proof. By assumption, u_i, u_i' and u_i'' in the scheme (2) are B-series $B(\psi_i, u_0, u_0', u_0'')$, $B(\frac{\rho}{h}\bar{\psi}_i, u_0, u_0', u_0'')$ and $B(\frac{\rho(\rho-1)}{h^2}\bar{\bar{\psi}}_i, u_0, u_0', u_0'')$ respectively, from Lemma 5 we have

$$h^4 F(u_i, u_i', u_i'') = B(\psi_i^{(4)}, u_0, u_0', u_0'') = \sum_{t \in RT \setminus \{\tau_1,\tau_2,\tau_3\}} \frac{h^{\rho(t)}}{\rho(t)!} \psi^{(4)}\, \gamma(t)\, \alpha(t)\, F(t)(u_0, u_0', u_0'', u_0''').$$

Therefore,

$$u_1(x_0+h) = u_0 + h u_0' + \frac{1}{2}h^2 u_0'' + \frac{1}{6}h^3 u_0''' + \sum_{i=1}^s b_i B(\psi_i^{(4)}, u_0, u_0', u_0'')$$

$$= u_0 + h u_0' + \frac{1}{2}h^2 u_0'' + \frac{1}{6}h^3 u_0''' + \sum_{t \in RT \setminus \{\tau_1,\tau_2,\tau_3\}} \frac{h^{\rho(t)}}{\rho(t)!} \varphi(t)\, \gamma(t)\, \alpha(t)\, F(t)(u_0, u_0', u_0'', u_0'''),$$

$$u_1'(x_0+h) = u_0' + h u_0'' + \frac{1}{2}h^2 u_0''' + \frac{1}{h}\sum_{i=1}^s b_i' B(\psi_i^{(4)}, u_0, u_0', u_0'')$$

$$= u_0' + h u_0'' + \frac{1}{2}h^2 u_0''' + \sum_{t \in RT \setminus \{\tau_1,\tau_2,\tau_3\}} \frac{h^{\rho(t)-1}}{\rho(t)!} \varphi'(t)\, \gamma(t)\, \alpha(t)\, F(t)(u_0, u_0', u_0'', u_0'''),$$

$$u_1''(x_0+h) = u_0'' + h u_0''' + \frac{1}{h^2}\sum_{i=1}^s b_i'' B(\psi_i^{(4)}, u_0, u_0', u_0'')$$

$$= u_0'' + h u_0''' + \sum_{t \in RT \setminus \{\tau_1,\tau_2,\tau_3\}} \frac{h^{\rho(t)-2}}{\rho(t)!} \varphi''(t)\, \gamma(t)\, \alpha(t)\, F(t)(u_0, u_0', u_0'', u_0'''),$$

$$u_1'''(x_0+h) = u_0''' + \frac{1}{h^3}\sum_{i=1}^s b_i''' B(\psi_i^{(4)}, u_0, u_0', u_0'')$$

$$= u_0''' + \sum_{t \in RT \setminus \{\tau_1,\tau_2,\tau_3\}} \frac{h^{\rho(t)-3}}{\rho(t)!} \varphi'''(t)\, \gamma(t)\, \alpha(t)\, F(t)(u_0, u_0', u_0'', u_0''').$$

□

3.4. Algebraic Order Conditions

Through Theorem 1 and 3, we arrived at the major goal of this study.

Theorem 4. *The RKTF method (2) has order $q(4 \leq q)$ if and only if the following conditions are satisfied as given in Hussain et al. [25])*

(i) $\varphi(t) = \frac{1}{\gamma(t)}$, $\rho(t) \leq q$,

(ii) $\varphi'(t) = \frac{\rho(t)}{\gamma(t)}$, $\rho(t) \leq q+1$,

(iii) $\varphi''(t) = \frac{\rho(t)(\rho(t)-1)}{\gamma(t)}$, $\rho(t) \leq q+2$,

(iv) $\varphi'''(t) = \frac{\rho(t)(\rho(t)-1)(\rho(t)-2)}{\gamma(t)}$, $\rho(t) \leq q+3$.

Corollary 1. *(see Hussain et al. [25]).* Assume that

$$t^* = [\tau_1^{\mu_1 + 2\mu_2 + 3\mu_3}, t_4^{\mu_4}, \ldots, t_r^{\mu_r}, < t_{r+1}^{\mu_{r+1}}, \ldots, t_n^{\mu_n} >, < t_{n+1}^{\mu_{n+1}}, \ldots, t_m^{\mu_m} >]_4,$$

$$\hat{t} = [\tau_1^{\mu_1}, \tau_2^{\mu_2}, \tau_3^{\mu_3}, t_4^{\mu_4}, \ldots, t_r^{\mu_r}, < t_{r+1}^{\mu_{r+1}}, \ldots, t_n^{\mu_n} >, < t_{n+1}^{\mu_{n+1}}, \ldots, t_m^{\mu_m} >]_4$$

where $t_4, \ldots, t_r$ are distinct and different from τ_1, τ_2 and τ_3 and $t_{r+1}, \ldots, t_n$, $t_{n+1}, \ldots, t_m$ are distinct. Then

$$\eta_i(\hat{t}) = \frac{1}{2^{\mu_2} 6^{\mu_3}} \eta_i(t^*), \quad \rho(\hat{t}) = \rho_i(t^*), \quad \gamma(\hat{t}) = 2^{\mu_2} 6^{\mu_3} \gamma(t^*).$$

Based on Corollary 1 assuming that the t^* and $\hat{t}$ trees grant the same order conditions, then these trees are equivalent. Thus, we can delete some trees since they are equivalent. For example, in Table 2 trees t_{61} and t_{68} of sixth-order are equivalent.

Based on Theorem 4 and Corollary 1, the algebraic order conditions up to order six for the RKTF formula can be presented as follows:

order 1:
$$b'''^T e = 1. \tag{22}$$

order 2:
$$b'''^T c = \frac{1}{2}, \quad b''^T e = \frac{1}{2}. \tag{23}$$

order 3:
$$b'''^T c^2 = \frac{1}{3}, \quad b'''^T \bar{\bar{A}} = \frac{1}{6}, \quad b''^T c = \frac{1}{6}, \quad b'^T e = \frac{1}{6}. \tag{24}$$

order 4:
$$b'''^T c^3 = \frac{1}{4}, \quad b'''^T \bar{A} = \frac{1}{24}, \quad b'''^T (c.\bar{A}e) = \frac{1}{8}, \quad b'''^T \bar{A}c = \frac{1}{24},$$
$$b''^T c^2 = \frac{1}{12}, \quad b''^T \bar{\bar{A}} = \frac{1}{24}, \quad b'^T c = \frac{1}{24}, \quad b^T e = \frac{1}{24}. \tag{25}$$

order 5:
$$b'''^T c^4 = \frac{1}{5}, \quad b'''^T A = \frac{1}{120}, \quad b'''^T \bar{A}c = \frac{1}{120}, \quad b'''^T c.\bar{A}e = \frac{1}{30}, \quad b'''^T \bar{A}c^2 = \frac{1}{60},$$
$$b'''^T (c^2.\bar{A}) = \frac{1}{10}, \quad b'''^T (c.\bar{A}c) = \frac{1}{30}, \quad b''^T c^3 = \frac{1}{20}, \quad b''^T \bar{A} = \frac{1}{120}, \quad b''^T (c.\bar{A}e) = \frac{1}{40},$$
$$b''^T \bar{A}c = \frac{1}{120}, \quad b'^T c^2 = \frac{1}{60}, \quad b'^T \bar{\bar{A}} = \frac{1}{120}, \quad b^T c = \frac{1}{120}. \tag{26}$$

order 6:

$$b'''^T c^5 = \frac{1}{6}, \quad b'''^T Ac = \frac{1}{720}, \quad b'''^T(c.Ae) = \frac{1}{144}, \quad b'''^T \bar{A}c^2 = \frac{1}{360},$$

$$b'''^T(c.\bar{A}c^2) = \frac{1}{72}, \quad b'''^T(\bar{A}c^2) = \frac{1}{360}, \quad b'''^T(c^2.\bar{A}e) = \frac{1}{36},$$

$$b'''^T(c.\bar{A}c) = \frac{1}{144}, \quad b'''^T(c^3.\bar{A}e) = \frac{1}{12}, \quad b'''^T \bar{A}c^3 = \frac{1}{120},$$

$$b'''^T(c.\bar{A}c^2) = \frac{1}{144}, \quad b'''^T(c^2.\bar{A}c) = \frac{1}{36}, \quad b''^T(c.\bar{A}c) = \frac{1}{180}, \quad b''^T c^4 = \frac{1}{30},$$

$$b''^T A = \frac{1}{720}, \quad b''^T \bar{A}c = \frac{1}{720}, \quad b''^T(c.\bar{A}e) = \frac{1}{180}, \quad b''^T \bar{A}c^2 = \frac{1}{360},$$

$$b''^T(c^2.\bar{A}e) = \frac{1}{60}, \quad b'^T c^3 = \frac{1}{120}, \quad b'^T \bar{A} = \frac{1}{720}, \quad b'^T(c.\bar{A}e) = \frac{1}{240},$$

$$b'^T \bar{A}c = \frac{1}{720}, \quad b^T \bar{A} = \frac{1}{720}, \quad b^T c^2 = \frac{1}{360}. \tag{27}$$

The following simplifying assumption is used to reduce the number of equations to be solved: $\sum \bar{a}_{ij} = \frac{c_i^2}{2}$.

3.5. Zero-Stability of the New Method

Here, we will discuss the zero-stability of the new techniques. It is stable at zero significance to prove the convergence of multi-step techniques and stability (see [10,11]). In [29], also discussed on the zero-stability to obtain the upper boundedness of the multi-steps methods. Now, the first characteristic polynomial for the RKTF method for Equation (2) is based on the following equation:

$$\begin{bmatrix} 1 & 0 & 0 & 0 \\ 0 & 1 & 0 & 0 \\ 0 & 0 & 1 & 0 \\ 0 & 0 & 0 & 1 \end{bmatrix} \begin{bmatrix} u_{n+1} \\ hu'_{n+1} \\ h^2 u''_{n+1} \\ h^3 u'''_{n+1} \end{bmatrix} = \begin{bmatrix} 1 & 1 & \frac{1}{2} & \frac{1}{6} \\ 0 & 1 & 1 & \frac{1}{2} \\ 0 & 0 & 1 & 1 \\ 0 & 0 & 0 & 1 \end{bmatrix} \begin{bmatrix} u_n \\ hu'_n \\ h^2 u''_n \\ h^3 u'''_n \end{bmatrix},$$

where $I = \begin{bmatrix} 1 & 0 & 0 & 0 \\ 0 & 1 & 0 & 0 \\ 0 & 0 & 1 & 0 \\ 0 & 0 & 0 & 1 \end{bmatrix}$ is the identity matrix coefficient of $u_{n+1}, hu'_{n+1}, h^2 u''_{n+1}$ and $h^3 u'''_{n+1}$

and $A = \begin{bmatrix} 1 & 1 & \frac{1}{2} & \frac{1}{6} \\ 0 & 1 & 1 & \frac{1}{2} \\ 0 & 0 & 1 & 1 \\ 0 & 0 & 0 & 1 \end{bmatrix}$ is matrix coefficient of $u_n, hu'_n, h^2 u''_n$ and $h^3 u'''_n$, respectively.

Then, the first characteristic polynomial of new methods is

$$\rho(\zeta) = \det[I\zeta - A] = \begin{vmatrix} \zeta - 1 & -1 & -\frac{1}{2} & -\frac{1}{6} \\ 0 & \zeta - 1 & -1 & -\frac{1}{2} \\ 0 & 0 & \zeta - 1 & -1 \\ 0 & 0 & 0 & \zeta - 1 \end{vmatrix}.$$

thus, $\rho(\zeta) = (\zeta - 1)^4$. By solving the characteristic polynomial, we obtain the roots, $\zeta = 1, 1, 1, 1$. Therefore, the RKTF methods is zero stable since the roots of the characteristic polynomial have modulus less than or equal to one. The RKTF is consistent because the RKTF has order $p \geq 4$. This property, with the zero stable of the methods, implies the convergence of the RKT method.

4. Construction of the RKTF Methods

According the order conditions stated in Section 3.4 before we proceed to construct explicit RKTF methods. The local truncated error for the q order RKTF technique is defined as follows:

$$\| L_g^{(q+1)} \|_2 = \left(\sum_{i=1}^{n_{q+1}} \left(L_i^{(q+1)} \right)^2 + \sum_{i=1}^{n'_{q+1}} \left(L_i'^{(q+1)} \right)^2 + \sum_{i=1}^{n''_{q+1}} \left(L_i''^{(q+1)} \right)^2 + \sum_{i=1}^{n'''_{q+1}} \left(L_i'''^{(q+1)} \right)^2 \right)^{\frac{1}{2}} \quad (28)$$

where $L^{(q+1)}, L'^{(q+1)}, L''^{(q+1)}$ and $L'''^{(q+1)}$ are the local truncation error terms for u, u', u'' and u''' respectively, $L_g^{(q+1)}$ is the global local truncation error.

4.1. A Three-Stage Fourth-Order RKTF Method

In this subsection the derivation of the three-stage RKTF technique of order four by using the algebraic order conditions up to order four and simplifying assumption $\sum \bar{a}_{ij} = \frac{c_i^2}{2}$ will be considered. The resulting system consists of 15 nonlinear equations with 23 unknown variables, solving the system simultaneously and the family of solution in terms of $a_{21}, a_{31}, a_{32}, \bar{a}_{32}, b_2, c_3$ and letting $\bar{a}_{21} = 0, b_3 = 0$, and $b'_3 = 0$ are given as follows:

$$\bar{a}_{31} = -\bar{a}_{32} + \frac{3}{4}c_3 - 2c_3^2 + \frac{3}{2}c_3^3,$$

$$\bar{a}_{21} = \frac{(-3 + 4c_3)^2}{8(-2 + 3c_3)^2}, \quad \bar{a}_{31} = -\frac{c_3(14c_3 - 20c_3^2 - 3 + 9c_3^3)}{-3 + 4c_3},$$

$$\bar{a}_{32} = \frac{(3 - 8c_3 + 6c_3^2)c_3(-2 + 3c_3)}{2(-3 + 4c_3)}, \quad b_1 = \frac{1}{24} - b_2, \quad b'_1 = \frac{-4 + 5c_3}{12(-3 + 4c_3)},$$

$$b'_2 = \frac{-2 + 3c_3}{12(-3 + 4c_3)}, \quad b''_1 = \frac{6c_3^2 - 6c_3 + 1}{6(-3 + 4c_3)c_3}, \quad b''_2 = \frac{(2 - 7c_3 + 6c_3^2)(-2 + 3c_3)}{3(3 - 8c_3 + 6c_3^2)(-3 + 4c_3)},$$

$$b''_3 = \frac{-(-1 + c_3)}{6(3 - 8c_3 + 6c_3^2)c_3}, \quad b'''_1 = \frac{6c_3^2 - 6c_3 + 1}{6(-3 + 4c_3)c_3}, \quad b'''_2 = \frac{2(4 - 12c_3 + 9c_3^2)(-2 + 3c_3)}{3(3 - 8c_3 + 6c_3^2)(-3 + 4c_3)},$$

$$b'''_3 = \frac{1}{6(3 - 8c_3 + 6c_3^2)c_3}, \quad c_2 = \frac{-3 + 4c_3}{2(-2 + 3c_3)}.$$

Next, we minimize the truncation error term by using minimize command in Maple. Thus, for the optimized value of coefficients in fractional form we chose $a_{21} = -\frac{23}{50}$, $a_{31} = \frac{8}{25}$, $a_{32} = \frac{8}{25}$, $\bar{a}_{32} = \frac{3}{50}$, $c_3 = \frac{21}{25}$ and $b_2 = \frac{1}{50}$ with these values $\| \tau_g^{(5)} \|_2 = 7.98593 \times 10^{-3}$. Finally, all the parameters of three-stage fourth-order RKTF approach that will be denoted as RKTF4 can be written as follows (see Table 3):

Table 3. The RKTF4 Method.

0	0			0			0				
$\frac{9}{26}$	$-\frac{23}{50}$	0		0	0		$\frac{81}{1352}$	0			
$\frac{21}{25}$	$\frac{8}{25}$	$\frac{8}{25}$	0	$\frac{2991}{62,500}$	$\frac{3}{50}$	0	$\frac{644}{15,625}$	$\frac{9737}{31,250}$	0		
	$\frac{13}{600}$	$\frac{1}{50}$	0	$\frac{5}{108}$	$\frac{13}{108}$	0	$\frac{121}{1134}$	$\frac{2873}{8667}$	$\frac{1250}{20223}$	$\frac{121}{1134}$	$\frac{4394}{8667}$	$\frac{15,625}{40,446}$

4.2. A Four-Stage RKTF Method of Order Five

For four-stage RKTF technique of order five, the algebraic conditions up to order five will be solved. The resulting system consists of 29 nonlinear equations with 37 unknown variables, solving the system together will give a family of solution with 11 free parameters of $a_{21}, a_{32}, a_{42}, a_{43}, \bar{a}_{21}, \bar{a}_{42}, \bar{a}_{43}, b'_4, c_2, b_3$ and b_4 are given as follows:

$$\bar{a}_{31} = \frac{1}{10(16c_2^2 - 8c_2 + 1)(5c_2 - 4)(4c_2 - 1)^3}(50000 c_2^6 \bar{a}_{4,2} - 880 c_2^5 - 1040 c_2^2 \bar{a}_{2,1} - 18750 c_2^3 \bar{a}_{4,2}$$
$$+ 11250 c_2^3 \bar{a}_{4,3} + 800 c_2^3 \bar{a}_{2,1} + 66250 c_2^4 \bar{a}_{4,2} - 21250 c_2^4 \bar{a}_{4,3} - 97500 c_2^5 \bar{a}_{4,2} + 12500 c_2^5 \bar{a}_{4,3}$$
$$- 40 \bar{a}_{2,1} + 320 c_2^2 - 32 c_2 - 1134 c_2^3 + 1684 c_2^4 + 1875 c_2^2 \bar{a}_{4,2} - 1875 c_2^2 \bar{a}_{4,3} + 370 c_2 \bar{a}_{2,1}),$$

$$\bar{a}_{21} = \frac{c_2^2}{2}, \bar{a}_{31} = \frac{c_2^2}{2(4c_2-1)^2}, \bar{a}_{41} = -\frac{4(50c_2^4 - 260c_2^3 + 321c_2^2 - 128c_2 + 16)}{625 c_2^2 (10 c_2^2 - 12 c_2 + 3)},$$

$$a_{31} = \frac{-1}{10(16c_2^2 - 8c_2 + 1)(5c_2 - 4)(4c_2 - 1)^2}(12500 a_{4,2} c_2^5 + 12500 a_{4,3} c_2^5 - 220 c_2^5 + 12800 a_{3,2} c_2^5$$
$$- 23040 a_{3,2} c_2^4 - 21250 a_{4,2} c_2^4 - 21250 a_{4,3} c_2^4 + 366 c_2^4 - 192 c_2^3 + 15040 a_{3,2} c_2^3 + 11250 a_{4,2} c_2^3$$
$$+ 11250 a_{4,3} c_2^3 - 1875 a_{4,2} c_2^2 - 1875 a_{4,3} c_2^2 + 32 c_2^2 - 4640 a_{3,2} c_2^2 + 690 a_{3,2} c_2 + 110 a_{2,1} c_2 - 40 a_{3,2}$$
$$- 40 a_{2,1}),$$

$$\bar{a}_{32} = \frac{-(2c_2 - 1)c_2}{10(16c_2^2 - 8c_2 + 1)(5c_2 - 4)(4c_2 - 1)^3}(-440 c_2^3 + 622 c_2^2 - 256 c_2 + 32 + 25000 c_2^4 \bar{a}_{4,2}$$
$$- 36250 c_2^3 \bar{a}_{4,2} + 15000 c_2^2 \bar{a}_{4,2} - 1875 \bar{a}_{4,2} c_2 + 6250 c_2^3 \bar{a}_{4,3} - 7500 c_2^2 \bar{a}_{4,3} + 1875 c_2 \bar{a}_{4,3}),$$

$$\bar{a}_{41} = -\frac{1}{125 c_2 (10 c_2^2 - 12 c_2 + 3)}(1250 c_2^3 \bar{a}_{4,3} - 110 c_2^3 + 1250 c_2^3 \bar{a}_{4,2} - 1500 c_2^2 \bar{a}_{4,3} + 128 c_2^2$$
$$- 1500 c_2^2 \bar{a}_{4,2} - 32 c_2 + 375 c_2 \bar{a}_{4,3} + 375 \bar{a}_{4,2} c_2 + 20 \bar{a}_{2,1}),$$

$$\bar{a}_{42} = \frac{(5c_2 - 4)(33c_2^2 - 34c_2 + 8)}{625(2c_2 - 1)c_2^2 (10c_2^2 - 12c_2 + 3)}, \bar{a}_{43} = \frac{(4c_2 - 1)^2 (275 c_2^3 - 430 c_2^2 + 208 c_2 - 32)}{625(2c_2 - 1)c_2^2 (10c_2^2 - 12c_2 + 3)},$$

$$b_1 = \frac{660 c_2^2 b_4 + 20 c_2 - 768 c_2 b_4 - 5 + 192 b_4}{300 c_2^2}, b_2 = -\frac{-15 c_2 + 1056 c_2 b_4 - 384 b_4 + 10}{1200(2c_2 - 1) c_2^2}, c_4 = \frac{4}{5},$$

$$b'_1 = -\frac{-20 c_2^2 + 9 c_2 - 240 b'_3 c_2 - 504 b'_4 c_2 + 96 b'_4 - 1 + 480 b'_3 c_2^2 + 480 b'_4 c_2^2}{120 c_2 (4c_2 - 1)}, c_3 = \frac{c_2}{4c_2 - 1},$$

$$b'_2 = -\frac{120 b'_3 c_2 + 384 b'_4 c_2 - 96 b'_4 - 4 c_2 + 1}{120 c_2 (4c_2 - 1)}, b''_1 = \frac{2c_2^2 + 4c_2 - 1}{48 c_2^2}, b''_2 = -\frac{c_2 - 1}{24 c_2^2 (5c_2 - 4)(2c_2 - 1)},$$

$$b''_3 = \frac{192 c_2^4 - 208 c_2^3 + 84 c_2^2 - 15 c_2 + 1}{24(2c_2 - 1) c_2^2 (11 c_2 - 4)}, b''_4 = \frac{25(10 c_2^2 - 12 c_2 + 3)}{48(11 c_2 - 4)(5 c_2 - 4)}, b'''_1 = \frac{2c_2^2 + 4c_2 - 1}{48 c_2^2},$$

$$b'''_2 = \frac{1}{24 c_2^2 (5c_2 - 4)(2c_2 - 1)}, b'''_3 = \frac{(4c_2 - 1)^2 (16 c_2^2 - 8 c_2 + 1)}{24 c_2^2 (11 c_2 - 4)(2 c_2 - 1)}, b'''_4 = \frac{125(10 c_2^2 - 12 c_2 + 3)}{48(11 c_2 - 4)(5 c_2 - 4)}.$$

Minimizing the local truncation error norms and the optimized value of coefficients in fractional form will result in $a_{21} = \frac{1}{2}$, $a_{32} = -\frac{1}{25}$, $a_{42} = -\frac{6}{25}$, $a_{43} = \frac{13}{25}$, $\bar{a}_{21} = \frac{1}{200}$, $\bar{a}_{42} = \frac{1}{1000}$, $\bar{a}_{43} = \frac{3}{100}$, $c_2 = \frac{37}{50}$, $b_3 = \frac{2}{5}$, $b_4 = \frac{11}{10}$ and $b'_4 = \frac{3}{100}$ with these values $\| \tau_g^{(6)} \|_2 = 8.771395898 \times 10^{-3}$.

Lastly, all the parameters of four-stage fifth-order RKTF method indicated by RKTF5 can be written as follows :

$$c_2 = \frac{37}{50}, a_{21} = \frac{1}{2}, \bar{a}_{21} = \frac{1}{200}, \bar{\bar{a}}_{21} = \frac{1369}{5000}, c_3 = \frac{37}{98}, a_{31} = \frac{29,560,597}{288,240,050}, a_{32} = -\frac{1}{25},$$

$$\bar{a}_{31} = \frac{23,408,341}{4,519,603,984}, \bar{a}_{32} = -\frac{20,407,091}{4,519,603,984}, \bar{\bar{a}}_{31} = \frac{1369}{19,208}, \bar{\bar{a}}_{32} = 0, c_4 = \frac{4}{5}, a_{41} = 0, a_{42} = -\frac{6}{25},$$

$$a_{43} = \frac{13}{25}, \bar{a}_{41} = \frac{77,969}{3,737,000}, \bar{a}_{42} = \frac{1}{1000}, \bar{a}_{43} = \frac{3}{100}, \bar{\bar{a}}_{41} = \frac{3,347,324}{17,283,625}, \bar{\bar{a}}_{42} = \frac{2277}{553,076},$$

$$\bar{\bar{a}}_{43} = \frac{8,449,119}{69,134,500}, b_1 = -\frac{1107}{14,504}, b_2 = -\frac{30,067}{21,756}, b_3 = \frac{2}{5}, b_4 = \frac{11}{10}, b_1' = \frac{116,911}{2,053,500}, b_2' = -\frac{13,529}{394,272},$$

$$b_3' = \frac{5,620,741}{49,284,000}, b_4' = \frac{3}{100}, b_1'' = \frac{1273}{10,952}, b_2'' = -\frac{40,625}{295,704}, b_3'' = \frac{7,176,589}{20,403,576}, b_4'' = \frac{2525}{14,904},$$

$$b_1''' = \frac{1273}{10,952}, b_2''' = -\frac{78,125}{147,852}, b_3''' = \frac{5,764,801}{10,201,788}, b_4''' = \frac{12,625}{14,904}$$

5. Numerical Experiments

Some of the problems involving $u^{(4)} = f(x, u, u', u'')$ are tested in this section. The numerical results are compared with the results obtained when the same group of examples is transformed to a system of first order and is solved using the existing RK of the same order.

- RKTF5: the explicit RKTF method of order five with four-stage derived in this paper.
- RKTF4: the explicit RKTF method of order four with three-stage constructed in this paper.
- RKF5: the fifth-order RK method with six-stage given in Lambert [11].
- DOPRI5: the fifth-order RK method with seven-stage derived in Dormand [10].
- RK4: the classical RK method of order four with four-stage as given in Butcher [29].
- RKM4: the RK method of order four with five-stage derived in Hairer [9].

Problem 1: (Linear System Inhomogeneous)

$$u_1^{(4)}(x) = -u_2''(x), \quad u_1(0) = 1, \quad u_1'(0) = 1, \quad u_1''(0) = 1, \quad u_1'''(0) = 1,$$
$$u_2^{(4)}(x) = -u_1''(x), \quad u_2(0) = -1, \quad u_2'(0) = -1, \quad u_2''(0) = -1, \quad u_2'''(0) = -1,$$
$$u_3^{(4)}(x) = -u_3''(x) - u_3(x) - \cos(x), \quad u_3(0) = -1, \quad u_3'(0) = 0, \quad u_3''(0) = 1, \quad u_3'''(0) = 0,$$
$$u_4^{(4)}(x) = -u_4''(x) - u_4(x) - 2\cos(x), \quad u_4(0) = -2, \quad u_4'(0) = 0, \quad u_4''(0) = 2, \quad u_4'''(0) = 0,$$

The exact solution is

$$u_1(x) = e^{(x)}, \quad u_2(x) = -e^{(x)}, \quad u_3(x) = -\cos(x), \quad u_4(x) = -2\cos(x),$$

Problem 2: (Homogeneous Linear Problem)

$$u^{(4)}(x) = -u''(x), \quad u(0) = 1, \quad u'(0) = 0, \quad u''(0) = -1, \quad u'''(0) = 0,$$

The exact solution is

$$u(x) = \cos(x).$$

Problem 3: (Inhomogeneous Nonlinear Problem)

$$u^{(4)}(x) = u^2(x) + \cos^2(x) - u''(x) - 1,$$
$$u(0) = 0, \quad u'(0) = 1, \quad u''(0) = 0, \quad u'''(0) = -1,$$

The exact solution is $u(x) = \sin(x)$.
Problem 4: (Inhomogeneous Linear Problem)

$$u^{(4)}(x) = -2u''(x) - u(x) + 1,$$
$$u(0) = 0, \quad u'(0) = 0, \quad u''(0) = 1, \quad u'''(0) = 0,$$

The exact solution is $u(x) = 1 - \cos(x)$.
Problem 5: Linear system homogeneous given in Hussain et al. [25]

$$u_1^{(4)}(x) = e^{3x} u_4(x), \qquad u_1(0) = 1, \quad u_1'(0) = -1, \quad u_1''(0) = 1, \quad u_1'''(0) = -1,$$
$$u_2^{(4)}(x) = 16 e^{-x} u_1(x), \qquad u_2(0) = 1, \quad u_2'(0) = -2, \quad u_2''(0) = 4, \quad u_2'''(0) = -8,$$
$$u_3^{(4)}(x) = 81 e^{-x} u_2(x), \qquad u_3(0) = 1, \quad u_3'(0) = -3, \quad u_3''(0) = 9, \quad u_3'''(0) = -27,$$
$$u_4^{(4)}(x) = 256 e^{-x} u_3(x), \qquad u_4(0) = 1, \quad u_4'(0) = -4, \quad u_4''(0) = 16, \quad u_4'''(0) = -64,$$

The exact solution is given by

$$u_1(x) = e^{-x}, \quad u_2(x) = e^{-2x}, \quad u_3(x) = e^{-3x}, \quad u_4(x) = e^{-4x}, 0 \leq x \leq 3.$$

Table 4. Numerical results for Problem 1 for RKTF4 method.

h	Methods	F.N	MAXE	TIME
	RKTF4	404	1.222871(−1)	0.017
0.1	RK4	1616	1.885232(−1)	0.037
	RKM4	2020	3.403273(−2)	0.065
	RKTF4	1600	1.338047(−4)	0.018
0.025	RK4	6400	7.022302(−4)	0.060
	RKM4	8000	1.194765(−4)	0.066
	RKTF4	3204	5.157453(−6)	0.019
0.0125	RK4	12,816	4.496182(−5)	0.064
	RKM4	16,020	7.571023(−6)	0.075
	RKTF4	6404	2.224660(−7)	0.020
0.00625	RK4	25,616	2.798824(−6)	0.068
	RKM4	32,020	4.633184(−6)	0.090

Table 5. Numerical results for Problem 2 for RKTF4 method.

h	Methods	F.N	MAXE	TIME
	RKTF4	12,000	5.534239(−5)	0.020
0.1	RK4	64,000	6.414194(−4)	0.022
	RKM4	80,000	5.560571(−5)	0.039
	RKTF4	48,003	2.162515(−7)	0.025
0.025	RK4	256,016	2.365790(−6)	0.029
	RKM4	320,020	2.164114(−7)	0.041
	RKTF4	96,003	1.329278(−8)	0.026
0.0125	RK4	512,016	8.094855(−8)	0.044
	RKM4	640,020	1.330367(−8)	0.056
	RKTF4	192,000	1.193586(−9)	0.039
0.00625	RK4	1,024,000	5.429163(−9)	0.057
	RKM4	1,201,354	1.201354(−9)	0.063

Table 6. Numerical results for Problem 3 for RKTF4 method.

h	Methods	F.N	MAXE	TIME
0.1	RKTF4	303	5.505858(−5)	0.016
	RK4	1616	1.231418(−4)	0.018
	RKM4	2020	7.157474(−5)	0.019
0.025	RKTF4	1200	8.246706(−7)	0.018
	RK4	6400	4.384085(−7)	0.019
	RKM4	8000	2.778406(−7)	0.020
0.0125	RKTF4	2403	5.811466(−8)	0.020
	RK4	12,816	2.730099(−8)	0.022
	RKM4	16,020	1.765267(−8)	0.024
0.00625	RKTF4	4803	3.800168(−9)	0.021
	RK4	25,616	1.687264(−9)	0.025
	RKM4	32,020	1.102847(−9)	0.029

Table 7. Numerical results for Problem 4 for RKTF4 method.

h	Methods	F.N	MAXE	TIME
0.1	RKTF4	33	2.916673(−7)	0.013
	RK4	176	5.134405(−7)	0.015
	RKM4	220	7.799860(−8)	0.019
0.025	RKTF4	120	4.476108(−10)	0.025
	RK4	640	1.870891(−9)	0.029
	RKM4	800	3.044243(−10)	0.032
0.0125	RKTF4	243	2.326739(−11)	0.028
	RK4	1296	1.155365(−11)	0.033
	RKM4	1620	1.902623(−11)	0.057
0.00625	RKTF4	483	1.281475(−12)	0.039
	RK4	2576	7.177037(−12)	0.047
	RKM4	3220	1.187606(−12)	0.065

Table 8. Numerical results for Problem 5 for RKTF4 method.

h	Methods	F.N	MAXE	TIME
0.1	RKTF4	90	1.950979(0)	0.018
	RK4	480	3.529526(1)	0.022
	RKM4	600	8.144031(0)	0.025
0.025	RKTF4	363	1.141631(−3)	0.019
	RK4	1936	1.560395(−1)	0.026
	RKM4	2420	3.606455(−2)	0.038
0.0125	RKTF4	720	7.384678(−5)	0.021
	RK4	3840	8.711749(−3)	0.036
	RKM4	4800	2.014647(−3)	0.056
0.00625	RKTF4	1440	1.991337(−6)	0.024
	RK4	7680	5.445548(−4)	0.057
	RKM4	9600	1.259457(−4)	0.071

Table 9. Numerical results for Problem 1 for RKTF5 method.

h	Methods	F.N	MAXE	TIME
0.1	RKTF5	404	5.327998(−5)	0.021
	RKF5	2424	2.064967(−3)	0.024
	DOPRI5	2828	5.732670(−4)	0.027
0.025	RKTF5	1600	4.254471(−7)	0.022
	RKF5	9600	1.917218(−6)	0.031
	DOPRI5	11,200	5.706643(−7)	0.033
0.0125	RKTF5	3204	1.786611(−8)	0.023
	RKF5	19,224	6.053233(−8)	0.040
	DOPRI5	22,428	1.931767(−8)	0.043
0.00625	RKTF5	6404	6.002665(−9)	0.028
	RKF5	38,424	4.878530(−9)	0.064
	DOPRI5	44,828	7.250492(−9)	0.075

Table 10. Numerical results for Problem 2 for RKTF5 method.

h	Methods	F.N	MAXE	TIME
0.1	RKTF5	16,000	8.041249(−6)	0.020
	RKF5	96,000	3.609465(−6)	0.023
	DOPRI5	112,000	1.108071(−6)	0.026
0.025	RKTF5	64,004	7.853954(−9)	0.028
	RKF5	384,024	3.523595(−9)	0.032
	DOPRI5	448,028	1.085847(−9)	0.045
0.0125	RKTF5	128,004	4.112761(−10)	0.035
	RKF5	768,024	2.510125(−10)	0.067
	DOPRI5	896,028	2.290347(−10)	0.075
0.00625	RKTF5	256,000	3.651384(−10)	0.043
	RKF5	1,536,000	3.557808(−10)	0.090
	DOPRI5	1,792,000	3.557941(−10)	0.105

Table 11. Numerical results for Problem 3 for RKTF5 method.

h	Methods	F.N	MAXE	TIME
0.1	RKTF5	404	5.997978(−5)	0.020
	RKF5	2424	9.761318(−5)	0.029
	DOPRI5	2828	2.837350(−5)	0.037
0.025	RKTF5	1600	1.024417(−8)	0.021
	RKF5	9600	6.841436(−8)	0.043
	DOPRI5	11,200	1.404994(−8)	0.053
0.0125	RKTF5	3204	1.413103(−9)	0.035
	RKF5	19,224	2.045084(−9)	0.046
	DOPRI5	22,428	3.752920(−9)	0.059
0.00625	RKTF5	6404	1.078057(−10)	0.042
	RKF5	38,424	4.854006(−11)	0.072
	DOPRI5	44,828	1.245892(−11)	0.080

Table 12. Numerical results for Problem 4 for RKTF5 method.

h	Methods	F.N	MAXE	TIME
0.1	RKTF5	404	2.812051(−7)	0.016
	RKF5	2424	1.813420(−6)	0.017
	DOPRI5	2828	2.251424(−8)	0.025
0.025	RKTF5	1600	2.432738(−10)	0.018
	RKF5	9600	1.767513(−10)	0.019
	DOPRI5	12,200	6.083276(−10)	0.021
0.0125	RKTF5	3204	7.153833(−12)	0.021
	RKF5	19,224	5.553996(−12)	0.023
	DOPRI5	22,428	1.871991(−12)	0.027
0.00625	RKTF5	6404	4.845013(−13)	0.025
	RKF5	38,424	2.333522(−13)	0.030
	DOPRI5	44,828	3.941292(−13)	0.038

Table 13. Numerical results for Problem 5 for RKTF5 method.

h	Methods	F.N	MAXE	TIME
0.1	RKTF5	120	1.534759(−1)	0.018
	RKF5	720	6.153184(−1)	0.023
	DOPRI5	840	5.531381(−1)	0.028
0.025	RKTF5	484	3.920592(−5)	0.021
	RKF5	2904	6.983629(−4)	0.034
	DOPRI5	3388	1.877126(−5)	0.039
0.0125	RKTF5	5760	1.960240(−5)	0.024
	RKF5	5760	1.960240(−5)	0.061
	DOPRI5	6720	3.310441(−6)	0.074
0.00625	RKTF5	1920	1.468912(−7)	0.030
	RKF5	11,520	6.140969(−7)	0.065
	DOPRI5	13,440	7.817222(−8)	0.121

6. Application to Problem from Ship Dynamics

This new technique is used to solve a physical problem from ship dynamics. As declared by Wu et al. [3], when a sinusoidal wave of hesitancy Ω passes along a ship or offshore structure, the resultant fluid actions vary with time x. In a specific status for the research by Wu et al. [3], the fourth-order problem is presented as

$$u^{(4)} = -3u'' - u(2 + \epsilon \cos(\Omega x)), \quad x > 0 \tag{29}$$

which is based on several initial conditions:

$$u(0) = 1, \quad u'(0) = u''(0) = u'''(0) = 0.$$

where $\epsilon = 0$ for the presence of the theoretical solution, $y(x) = 2\cos(x) - \cos(x\sqrt{2})$. The theoretical solution is indeterminate when $\epsilon \neq 0$ (see Twizell [4]). Previously, somewhat numerical experiences for solving ordinary differential equations of order four have been expanded to solve ship dynamics. Numerical realization was offered by Twizell [4] and Cortell [5] in connection with the order four ordinary differential Equation (29) when $\epsilon = 0$ and $\epsilon = 1$ for $\Omega = 0.25(\sqrt{2}-1)$. Instead of solving the order four ordinary differential equations directly, Twizell [4] and Cortell [5] opined that traditional path is alleviation way for first order ODEs. Twizell [4] constructed the global extrapolation with a family of numerical formulas to raise the order of the formulas. Furthermore, Cortell [5] developed the expansion of the classical Runge-Kutta formula.

Table 14. Numerical results for Problem (29) for RKTF4 method with $\epsilon = 0$.

h	Methods	F.N	MAXE	TIME
0.1	RKTF4	120	4.343559(−5)	0.016
	RK4	480	2.898981(−5)	0.017
	RKM4	600	4.708466(−5)	0.018
0.025	RKTF4	484	4.042540(−8)	0.018
	RK4	1936	1.106125(−7)	0.058
	RKM4	2420	1.828451(−8)	0.061
0.0125	RKTF4	960	1.560340(−9)	0.034
	RK4	3840	6.884182(−9)	0.063
	RKM4	4800	1.142450(−9)	0.069
0.00625	RKTF4	1920	7.905333(−11)	0.056
	RK4	7680	4.293583(−10)	0.068
	RKM4	9600	7.143930(−11)	0.074

Table 15. Numerical results for Problem (29) for RKTF5 method with $\epsilon = 0$.

h	Methods	F.N	MAXE	TIME
0.1	RKTF5	120	8.312096(−7)	0.014
	RKF5	720	5.273884(−7)	0.015
	DOPRI5	840	1.489234(−7)	0.018
0.025	RKTF5	484	2.413660(−10)	0.023
	RKF5	2904	5.506529(−10)	0.059
	DOPRI5	3388	1.660703(−10)	0.062
0.0125	RKTF5	960	6.902590(−12)	0.052
	RKF5	5760	1.690381(−11)	0.063
	DOPRI5	6720	5.136336(−12)	0.066
0.00625	RKTF5	1920	2.069456(−13)	0.061
	RKF5	11,520	5.315748(−13)	0.069
	DOPRI5	13,440	1.643130(−13)	0.077

Table 16. Numerical results for Problem (29) for RKTF4 method with $\epsilon = 1$.

h	Methods	F.N	MAXE	TIME
0.5	RKTF4	6	4.255906(−3)	0.017
	RK4	48	2.418471(−3)	0.018
	RKM4	60	6.260650(−4)	0.023
0.2	RKTF4	15	5.127330(−5)	0.023
	RK4	96	7.798540(−5)	0.025
	RKM4	120	1.423970(−5)	0.033
0.1	RKTF4	33	1.854900(−6)	0.026
	RK4	176	5.067700(−6)	0.044
	RKM4	220	8.710000(−6)	0.069
0.025	RKTF4	120	3.300000(−9)	0.056
	RK4	640	2.010000(−8)	0.055
	RKM4	800	3.300000(−9)	0.074

Table 17. Numerical results for Problem (29) for RKTF5 method with $\epsilon = 1$.

h	Methods	F.N	MAXE	TIME
0.5	RKTF5	8	2.511868(−2)	0.016
	RKF5	72	1.646000(−3)	0.017
	DOPRI5	84	8.809400(−4)	0.019
0.2	RKTF5	20	5.912040(−6)	0.018
	RKF5	144	9.133000(−7)	0.021
	DOPRI5	168	4.852000(−7)	0.029
0.1	RKTF5	44	7.086000(−8)	0.026
	RKF5	264	1.930000(−8)	0.050
	DOPRI5	308	9.400000(−8)	0.060
0.025	RKTF5	160	1.000000(−10)	0.036
	RKF5	960	1.000000(−10)	0.065
	DOPRI5	1120	1.000000(−10)	0.076

7. Discussion and Conclusions

In this work, we are focusing on the algebraic theory of order conditions of RKTF method in the form of $u^{(4)} = f(x, u, u', u'')$ to solve ODEs of order four directly. Depending on the idea and concepts of rooted trees used to solve first and second order ordinary differential equations, many researchers have presented the definitions and algebraic theories of order algebraic conditions that we can see in [29–31]. Moreover, [32,33] introduced the idea and concept of B-series theory that are dependent on algebraic order conditions.

In fact, the motivation of our new work in using the B-series to construct RKT formula based on the algebraic order conditions developed in the form of $u^{(4)} = f(x, u, u', u'')$ to solve directly ODEs of order four. Furthermore, we developed three-stage of order four and four-stage of order five known as RKTF4 and RKTF5 methods, respectively.

The numerical outcomes are tabulated in Tables 4–17 and plotted in Figures 2–8. Those figures show the proficiency curves when compared the new methods with RKTF5, DOPRI5, RK4 and RKM4 methods by the number of function evaluations and maximum global error. Figures 2 and 3, RKTF4 and RKTF5 methods outperform over RKTF5, DOPRI5, RK4 and RKM4 methods in terms number function evaluations. Next, Figure 4 displays the efficacy of the new methods for inhomogeneous nonlinear problem. In Figures 5 and 6, we can see that RKTF4 and RKTF5 approaches are the more efficient and accurate methods compared to the other existing RK methods. Figures 7 and 8 show that the new methods require less function evaluations than RKF5, DOPRI5, RK4 and RKM4 methods. This is because when Equation (29) is solved using RKTF5, DOPRI5, RK4 and RKM4 methods, it needs to be reduced to a system of first-order equations which is four times the dimension. From numerical results in all tables, we noticed that the proposed methods outperform existing RK methods in terms of time for all step size. From numerical results in all figures, we noticed that the number of function evaluations of RKTF4 and RKTF5 methods are less than number of function evaluations for other existing RK methods and they have shown that the new methods are more accurate and appropriate when solving fourth-order ODEs in the form of $u^{(4)} = f(x, u, u', u'')$.

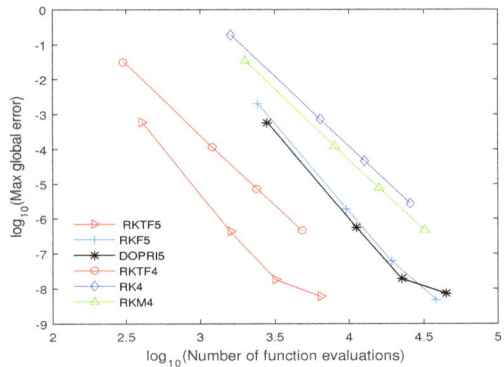

Figure 2. Efficiency curves for RKTF5, RKTF4, RKF5, DOPRI5, RK4 and RKM4 when solving Problem 1 with step size $h = 0.1, 0.025, 0.0125, 0.00625$ and $x_{end} = 10$.

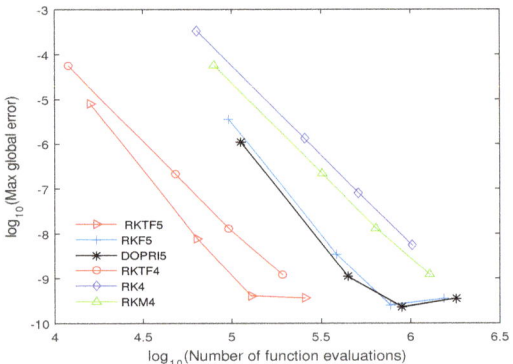

Figure 3. Efficiency curves for RKTF5, RKTF4, RKF5, DOPRI5, RK4 and RKM4 when solving Problem 2 with step size $h = 0.1, 0.025, 0.0125, 0.00625$ and $x_{end} = 400$.

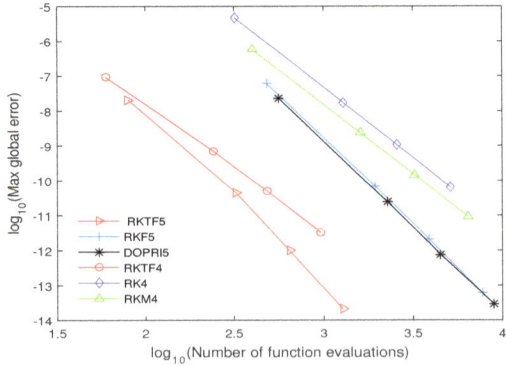

Figure 4. Efficiency curves for RKTF5, RKTF4, RKF5, DOPRI5, RK4 and RKM4 when solving Problem 3 with step size $h = 0.1, 0.025, 0.0125, 0.00625$ and $x_{end} = 10$.

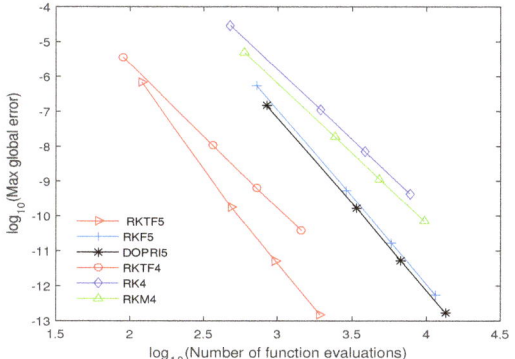

Figure 5. Efficiency curves for RKTF5, RKTF4, RKF5, DOPRI5, RK4 and RKM4 when solving Problem 4 with step size $h = 0.1, 0.025, 0.0125, 0.00625$ and $x_{end} = 10$.

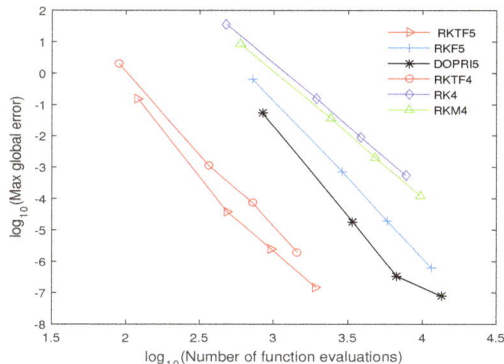

Figure 6. Efficiency curves for RKTF5, RKTF4, RKF5, DOPRI5, RK4 and RKM4 when solving Problem 5 with step size $h = 0.1, 0.025, 0.0125, 0.00625$ and $x_{end} = 3$.

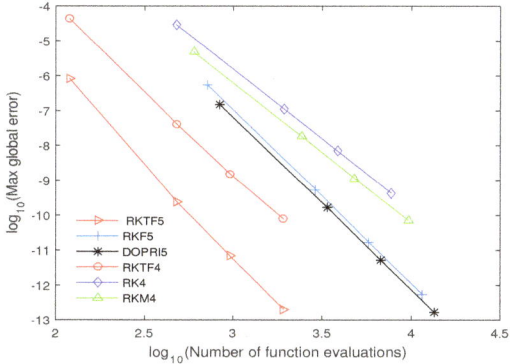

Figure 7. Efficiency curves for Equation (29) with $h = 0.1, 0.025, 0.0125, 0.00625$, $\epsilon = 0$ and $x_{end} = 3$.

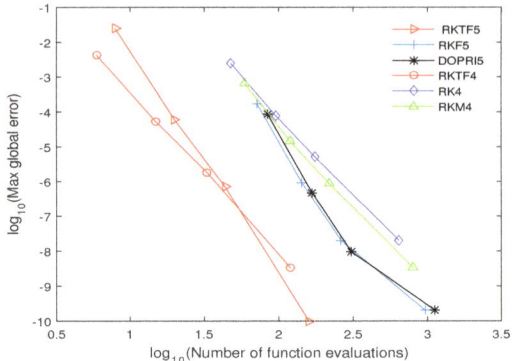

Figure 8. Efficiency curves for Equation (29) with $h = 0.5, 0.2, 0.1, 0.025$, $\epsilon = 1$ and $x_{end} = 1$.

Author Contributions: Conceptualization, N.G., F.A.F.; Methodology, N.G., F.A.F., N.S.; Formal Analysis, N.G., F.A.F.; Investigation, N.G., F.A.F.; Resources, N.G., F.A.F., N.S., F.I., Z.B.I.; Writing—Original Draft Preparation, N.G.; Writing—Review and Editing, N.S., F.I., Z.B.I.; Supervision, N.S.; Project Administration, N.S.; Funding Acquisition, UPM.

Funding: This study has been supported by Project Code: GP-IPS/2017/9526600 of the Universiti Putra Malaysia.

Acknowledgments: The authors are thankful to the referees for carefully reading the paper and for their valuable comments.

Conflicts of Interest: The authors declare that there is no conflict of interests regarding the publication of this paper.

Abbreviations

The following abbreviations are used in this manuscript:

h	Step size used.
IVPs	Initial value problems.
RKTF5	The explicit RKTF method of order five with four-stage derived in this paper.
RKTF4	The explicit RKTF method of order four with three-stage constructed in this paper.
RKF5	The fifth-order RK method with six-stage given in Lambert [11].
DOPRI5	The fifth-order RK method with seven-stage derived in Dormand [10].
RK4	The classical RK method of order four with four-stage as given in Butcher [29].
RKM4	The RK method of order four with five-stage derived in Hairer [9].

References

1. Malek, A.; Beidokhti, R.S. Numerical solution for high order differential equations using a hybrid neural network—Optimization method. *Appl. Math. Comput.* **2006**, *183*, 260–271. [CrossRef]
2. Alomari, A.; Anakira, N.R.; Bataineh, A.S.; Hashim, I. Approximate solution of nonlinear system of bvp arising in fluid flow problem. *Math. Probl. Eng.* **2013**, *2013*, 136043. [CrossRef]
3. Wu, X.; Wang, Y.; Price, W. Multiple resonances, responses, and parametric instabilities in offshore structures. *J. Ship Res.* **1988**, *32*, 285–296.
4. Twizell, E. A family of numerical methods for the solution of high-order general initial value problems. *Comput. Methods Appl. Mech. Eng.* **1988**, *67*, 15–25. [CrossRef]
5. Cortell, R. Application of the fourth-order Runge-Kutta method for the solution of high-order general initial value problems. *Comput. Struct.* **1993**, *49*, 897–900. [CrossRef]
6. Boutayeb, A.; Chetouani, A. A mini-review of numerical methods for high-order problems. *Int. J. Comput. Math.* **2007**, *84*, 563–579. [CrossRef]

7. Jator, S.N. Numerical integrators for fourth order initial and boundary value problems. *Int. J. Pure Appl. Math.* **2008**, *47*, 563–576.
8. Kelesoglu, O. The solution of fourth order boundary value problem arising out of the beam-column theory using adomian decomposition method. *Math. Probl. Eng.* **2014**, *2014*, 649471. [CrossRef]
9. Hairer, E.; Nrsett, G.; Wanner, G. *Solving Ordinary Differential Equations I: Nonstiff Problems*; Springer: Berlin, Germany, 1993.
10. Dormand, J.R. *Numerical Methods for Differential Equations: A Computational Approach*; CRC Press: Boca Raton, FL, USA, 1996; Volume 3.
11. Lambert, J.D. *Numerical Methods for Ordinary Differential Systems: The Initial Value Problem*; John Wiley & Sons, Inc.: New York, NY, USA, 1991.
12. Waeleh, N.; Majid, Z.; Ismail, F.; Suleiman, M. Numerical solution of higher order ordinary differential equations by direct block code. *J. Math. Stat.* **2012**, *8*, 77–81.
13. Awoyemi, D. Algorithmic collocation approach for direct solution of fourth-order initial-value problems of ordinary differential equations. *Int. J. Comput. Math.* **2005**, *82*, 321–329. [CrossRef]
14. Kayode, S.J. An efficient zero-stable numerical method for fourth-order differential equations. *Int. J. Math. Math. Sci.* **2008**, *2008*, 364021. [CrossRef]
15. Jator, S.N.; Li, J. A self-starting linear multistep method for a direct solution of the general second-order initial value problem. *Int. J. Comput. Math.* **2009**, *86*, 827–836. [CrossRef]
16. Jator, S.N. Solving second order initial value problems by a hybrid multistep method without predictors. *Appl. Math. Comput.* **2010**, *217*, 4036–4046. [CrossRef]
17. Waeleh, N.; Majid, Z.; Ismail, F. A new algorithm for solving higher order ivps of odes. *Appl. Math. Sci.* **2011**, *5*, 2795–2805.
18. Awoyemi, D. A p-stable linear multistep method for solving general third order ordinary differential equations. *Int. J. Comput. Math.* **2003**, *80*, 985–991. [CrossRef]
19. Awoyemi, D.; Idowu, O. A class of hybrid collocation methods for third-order ordinary differential equations. *Int. J. Comput. Math.* **2005**, *82*, 1287–1293. [CrossRef]
20. Ibrahim, Z.B.; Othman, K.I.; Suleiman, M. Implicit r-point block backward differentiation formula for solving first-order stiff odes. *Appl. Math. Comput.* **2007**, *186*, 558–565. [CrossRef]
21. Jain, M.; Iyengar, S.; Saldanha, J. Numerical solution of a fourth-order ordinary differential equation. *J. Eng. Math.* **1977**, *11*, 373–380. [CrossRef]
22. Mechee, M.; Senu, N.; Ismail, F.; Nikouravan, B.; Siri, Z. A three-stage fifth-order Runge-Kutta method for directly solving special third-order differential equation with application to thin film flow problem. *Math. Probl. Eng.* **2013**, *2013*, 795397. [CrossRef]
23. Mechee, M.; Ismail, F.; Siri, Z.; Senu, N. A four-stage sixth-order RKD method for directly solving special third-order ordinary differential equations. *Life Sci. J.* **2014**, *11*, 399–404.
24. Senu, N.; Mechee, M.; Ismail, F.; Siri, Z. Embedded explicit Runge–Kutta type methods for directly solving special third order differential equations $y''' = f(x,y)$. *Appl. Math. Comput.* **2014**, *240*, 281–293. [CrossRef]
25. Hussain, K.; Ismail, F.; Senu, N. Solving directly special fourth-order ordinary differential equations using Runge–Kutta type method. *J. Comput. Appl. Math.* **2016**, *306*, 179–199. [CrossRef]
26. Chen, Z.; Qiu, Z.; Li, J.; You, X. Two-derivative Runge-Kutta-Nyström methods for second-order ordinary differential equations. *Numer. Algorithms* **2015**, *70*, 897–927. [CrossRef]
27. You, X.; Chen, Z. Direct integrators of Runge–Kutta type for special third-order ordinary differential equations. *Appl. Numer. Math.* **2013**, *74*, 128–150. [CrossRef]
28. Hairer, E.; Lubich, C.; Wanner, G. *Geometric Numerical Integration: Structure-Preserving Algorithms for Ordinary Differential Equations*; Springer Science & Business Media: New York, NY, USA, 2006; Volume 31.
29. Butcher, J.C. *Numerical Methods for Ordinary Differential Equations*, 2nd ed.; John Wiley & Sons: Chichester, UK, 2008.
30. Butcher, J.C. Numerical methods for ordinary differential equations in the 20th century. *J. Comput. Appl. Math.* **2000**, *125*, 1–29. [CrossRef]

31. Butcher, J.C. An algebraic theory of integration methods. *Math. Comput.* **1972**, *26*, 79–106. [CrossRef]
32. Hairer, E.; Wanner, G. A theory for Nyström methods. *Numer. Math.* **1975**, *25*, 383–400. [CrossRef]
33. Hairer, E.; Wanner, G. On the butcher group and general multi-value methods. *Computing* **1974**, *13*, 1–15. [CrossRef]

© 2019 by the authors. Licensee MDPI, Basel, Switzerland. This article is an open access article distributed under the terms and conditions of the Creative Commons Attribution (CC BY) license (http://creativecommons.org/licenses/by/4.0/).

Article

Improvement of Risk Assessment Using Numerical Analysis for an Offshore Plant Dipole Antenna

Yun-Jeong Cho [1], Kichang Im [2], Dongkoo Shon [1], Daehoon Park [3] and Jong-Myon Kim [1],*

[1] School of Computer Engineering and Information Technology, University of Ulsan, Ulsan 44610, Korea; j_j7756@naver.com (Y.-J.C.); dongkoo88@gmail.com (D.S.)
[2] ICT Safety Convergence Center, University of Ulsan, Ulsan 44610, Korea; kichang@ulsan.ac.kr
[3] Convergence Technology Institute, Hyundai Heavy Industries Co., Ltd., Seongnam 13591, Korea; daehoon_park@hhi.co.kr
* Correspondence: jmkim07@ulsan.ac.kr; Tel.: +82-52-259-2217

Received: 30 October 2018; Accepted: 22 November 2018; Published: 1 December 2018

Abstract: This paper proposes a numerical analysis method for improving risk assessment of radio frequency (RF) hazards. To compare the results of conventional code analysis, the values required for dipole antenna risk assessment, which is widely used in offshore plants based on the British standards (BS) guide, are calculated using the proposed numerical analysis. Based on the BS (published document CENELEC technical report (PD CLC/TR) 50427:2004 and international electrotechnical commission (IEC) 60079 for an offshore plant dipole antenna, an initial assessment, a full assessment, and on-site test procedures are performed to determine if there is a potential risk of high-frequency ignition. Alternatively, numerical analysis is performed using the Ansys high frequency structure simulator (HFSS) tool to compare results based on the BS guide. The proposed method computes the effective field strength and power for the antenna without any special consideration of the structure to simplify the calculation. Experimental results show that the proposed numerical analysis outperforms the risk assessment based on the BS guide in accuracy of the evaluation.

Keywords: risk assessment; numerical analysis; ignition hazard; effective field strength; offshore plant

1. Introduction

At offshore plants, high-frequency waves, such as ultra high frequency (UHF) and very high frequency (VHF), are used for wireless communication, and automatic identification system (AIS), global positioning system (GPS), and radar scanners are installed. In such wireless communication devices, electromagnetic waves with high waves are generated. Additionally, structures such as metal objects, pipelines, crane ropes, etc. existing in the offshore plant can act as receivers. High-frequency electromagnetic waves from various devices induce voltage and current in metallic conductor structures at the offshore plant. The amplitude of such induced current depends on the wavelength of the transmitted signal, the surrounding electromagnetic field, and the shape and size of the structure. In addition, if the induced voltage or current is large enough, sparks can occur and cause large fires and explosions.

Offshore plants are subject to a variety of marine environmental conditions during operation, with more than 70% of accidents involving explosions or fires. This can lead to large-scale explosions in the event of an accident, leading to human casualties, and can cause serious marine pollution, which can lead to great economic and industrial losses. The possibility of fires and explosions caused by high-frequency radiation is analyzed considering the electromagnetic wave intensity generated from the communication facility of the offshore plant, the characteristics of the metallic structure acting as a receiving antenna, the size of the induced received power, and the characteristics of combustible material at the plant. The risk should be evaluated and reflected in design and construction.

The current state of related papers is as follows. Sang-Won Choi and Hyuk-Myun Kwon [1] performed an experiment to reduce electric shock and ignition by current and voltage induced by electromagnetic waves from a large crane. The necessity of various studies on the energy required to cause fires and explosions by a spark has been suggested. Eckhoff and Thomassen [2] studied various sources of ignition in offshore plants. Among them, the influence of high-frequency propagation was considered. According to their paper, electromagnetic waves are emitted by all systems that generate high-frequency electrical energy (10^4 Hz to 10^{11} Hz), and ignition sources can be generated if the field is strong enough and the receiver antenna is large enough. In this regard, a method of securing safety levels based on guidance was studied. Bradby [3] studied various practical applications of fire risk assessment based on BS 6656:2002 code. BS 6656:2002 [4] is a code detailing a systematic approach, such as initial assessment and full assessment, to assess radio frequency-induced ignition risk. The initial assessment is performed in three steps: (1) determine the size of the maximum vulnerable zone, (2) identify significant transmission sources within the vulnerable zone, and (3) screen each type of transmission source using BS 6656. Full assessment performs the full assessment methodology for the remaining sources. The study by Rajkumar and Bhattacharjee [5] carried out a step-by-step risk assessment from transmission based on BS 6656:2002. The evaluation step calculated the effective field strength according to frequency. The risk of fires and explosions in the explosion environment was evaluated by calculating the energy generated by sparks in the conductive structure caused by electromagnetic waves, and safety measures for the risk of radio frequency (RF) ignition were suggested. Wang [6] recommended RF mechanisms that directly or indirectly cause fires and explosions and summarized RF risk studies in flammable and explosive environments. In addition, some issues of RF risk studies were discussed, and RF risk studies were conducted.

Examining the status of related papers, most studies on high-frequency and RF risk assessment were based on the BS guide, and studies conducted in parallel with numerical analysis were rare. In addition, overall studies on land and sea have been conducted on the risk of high-frequency ignition. In this paper, a risk assessment for a dipole antenna commonly used in an offshore plant is performed. The assessment is based on BS PD CLC/TR 50427:2004 [7] and IEC 60079 [8–10] for risk assessment for high-frequency radiation. BS PD CLC/TR 50427:2004 is a guideline for assessing the risk of ignition of a facility where flammable gases due to RF emissions from communications, radar or other transmit antennas may be present. It specifies procedures and formulas necessary for risk assessment. IEC 60079 standard refers to IEC 60079-0, IEC 60079-10-1, and IEC 60079-10-2. IEC 60079-0 specifies the general requirements for the manufacture, testing and marking of electrical equipment and explosive (EX) components for use in explosive atmospheres. IEC 60079-10-1 is an international standard for identification and classification of explosive gas atmospheres and regions where combustible gases, vapors of mist can exist. IEC 60079-10-2 is an international standard for the identification and classification of explosive dust atmospheres and regions where combustible dust layers may be present. The IEC international standard was used to determine the gas threshold. At the same time, Ansys HFSS numerical analysis is performed to contribute to the accuracy of the existing risk assessment based on the BS guide.

2. Detailed Specification of the Analysis Object

In this paper, code and numerical analysis are performed by choosing antennas that are widely used in offshore plants. Code and numerical analysis are performed for both single and multiple transmission. In the code analysis, the risk of high-frequency ignition is evaluated according to the BS and IEC, and numerical analysis is compared with code analysis by modeling and analysis of the selected antenna.

A case study evaluates whether installed transmissions are at risk of high-frequency ignition in the vulnerable area. The case study is performed for a single transmission and multiple transmissions.

Additionally, the dipole transmission used selects MP4 among various AEP VHF antennas; the shape and detailed specifications for MP4 are shown in Table 1. Many marine companies use the MP4 [11] model because of the wide frequency bandwidth, the wide applicability, and the included mounting nut.

Table 1. Detailed specifications of MP4 [11].

Bottom Diameter (mm)	Length (m)	Weight (kg)	Frequency (MHz)	Max Input (W)	Gain (dB)	Impedance (Ω)
28	1.20	1.00	163	100	3	50

3. Risk Assessment Based on the BS Guide

Analysis of the risk assessment is performed in accordance with the BS guide, and methods for assessing potential RF ignition risk include initial assessment, full assessment, and on-site tests. There are two initial assessments: assessment of the risk to a particular plant and assessment of the risk from a particular transmitter. First, referring to the table of the BS guide for the risk to a particular plant, the different radii of vulnerable zones are calculated for all loop structures whose inside perimeter is 40 m or less, for horizontal loops whose height is 5 m or less, and for all other loop structures. The second is a risk assessment for a particular transmitter. For the cases where the inside circumference of the roof structure is less than 85 m at the frequency of 30 MHz or more and the maximum circumference of the largest structure in the plant is 40 m at the frequency of 30 MHz or less and the height is less than 5 m for the horizontal loop structure, the radius of the zone of vulnerable is calculated by referring to the table of BS guide. If the initial assessment indicates a risk inside the vulnerable zone, a full assessment is performed. In this paper, a risk assessment based on the BS guide was performed for a frequency of 30 MHz or more and an inner circumference of the loop structure of 85 m or less.

In the full assessment, the power or energy that can be extracted is calculated and compared with the threshold value according to the gas group. Finally, if a potential hazard is indicated in the full assessment, the power that can be extracted from the on-site test is measured and compared with the threshold value according to the gas group. If an on-site test indicates a hazard, measures such as plant design changes, plant movement, and transmission power reduction are considered. The full assessment of the entire procedure for performing a risk assessment from an antenna is shown in Figure 1, which is taken from the BS guide. As shown in Figure 1, all information about the plant or transmission is first collected (① of Figure 1), and then the effective field strength is calculated (② of Figure 1) according to the equation provided by the BS guide. Then, the equation is classified into three types according to frequency and polarization. Once the effective field strength is calculated, the extractable power is calculated (③ of Figure 1). The extractable power is calculated according to the formula provided by the BS guide, and classified into two types according to frequency. At this time, if the frequency is less than 30 MHz, the internal circumference of the structure is taken into account. If the frequency exceeds 30 MHz, the inside circumference of the structure is not considered. If the transmission used in the evaluation is a single transmission (④ of Figure 1), risk is evaluated by comparing it with the extracted power and the code-based threshold power. On the other hand, when the transmission is a multiple transmission (⑤ of Figure 1), the maximum extractable power is recalculated and compared with the threshold power to assess the risk. Finally, the threshold power is compared with the extractable power (⑥ of Figure 1). If the extractable power is smaller than the threshold power, it is checked whether there is no potential ignition risk (⑦ of Figure 1). On the other hand, if the extractable power is greater than or equal to the threshold power, it is checked whether there is a potential ignition risk (⑧ of Figure 1) and the on-site test should proceed.

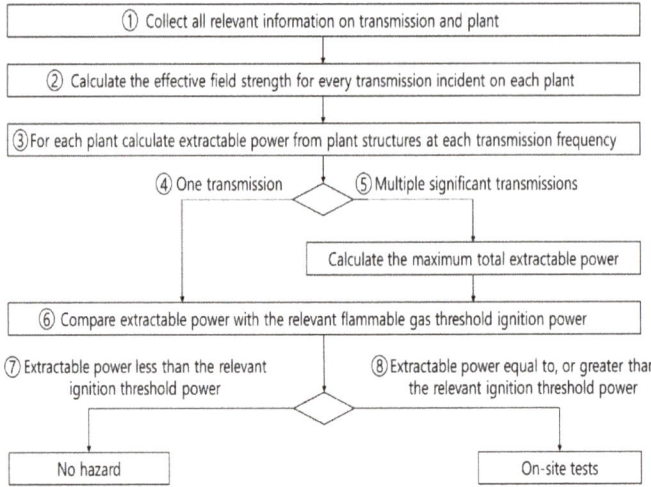

Figure 1. Full assessment procedure [7].

To evaluate the potential risks to the analysis object, risk assessment is performed based on the BS guide from Sections 3.1–3.3. For the dipole antenna to be analyzed in this paper, initial assessment is carried out for general and offshore plant-specific criteria in Section 3.1. As a result, of the analysis, it is judged if there is a potential risk, and the full assessment is performed in a single transmission environment in Section 3.2. Similarly, the full assessment in a multiple transmission environment is performed at Section 3.3. As a result, the analysis object determines whether there is a potential ignition risk under the condition.

3.1. Initial Assessment

Initial assessments are carried out on land and offshore plants in accordance with the BS guide. Offshore plants are a special concern in the assessment of ignition hazards by RF emissions. Therefore, the BS guide represents the standard for all transmissions, and the transmission criterion for offshore plants is a separate set of transmissions for special offshore plants.

A transmission in accordance with the BS guide is selected to determine the zone of vulnerability; then, it is determined if any gases or vapors that may be dangerous are inside or outside the zone of vulnerability. If gases or vapors are found to be outside the vulnerable area, the transmission is considered not to cause an RF ignition hazard, and the evaluation is stopped. Conversely, if gases or vapors are found to be located inside a vulnerable zone, the transmission is considered to cause an RF ignition hazard, and a full assessment procedure is followed.

First, initial assessment is conducted using general criteria, including land and sea. Here, the frequency is 30 MHz or more, and the loop structure is initialized to a specific transmission by setting the inner circumference to 85 m or less. Since the frequency of the antenna is 163 MHz, the transmission given in Table 2 is selected and performed. Table 2 is a reference to the radii of vulnerable zones of the BS guide and includes all land and sea transmissions. If the details of the transmissions are different from those shown in Table 2, the IIC group representing the largest vulnerable area of the closest equivalent transmission in Table 2 is selected. The initial assessment results are most similar to serial number 54 VHF and UHF land, fixed, and mobile and maritime mobile transmission, which is the largest IIC group, and the size of the vulnerable zone was determined to be 6 m.

Table 2. Radii of vulnerable zones (subsection) [7].

Serial No.	Type of Transmission	Frequency	Power	Modulation	Antenna Gain (dBi)	Radii of Vulnerable Zones (m)		
						Group I/IIA	Group IIB	Group IIC
54	VHF and UHF land, fixed and mobile and maritime mobile	68 MHz to 470 MHz	25 W	AM/RM	2	3.5	4.5	6

Second, offshore plants are a special issue in the assessment of risk of ignition by RF radiation. Thus, from general RF transmissions and antennas used in offshore plants, the vulnerable zone radius is based on the most common size of structures or cranes that can exist in an offshore plant according to BS PD CLC/TR 50427:2004. Antenna specifications and structural conditions apply equally to the general standards. Initial assessment should be performed by selecting the transmission given in Table 3. If the details of the transmission are not identical to those shown in Table 3, the vulnerable area of the nearest equivalent transmission is selected. The initial assessment results are compared to the transmission details given in Table 3. The most similar is marine VHF fixed, and the size of the vulnerable zone is determined to be 1.1 m.

Table 3. Radio frequency transmitters offshore (subsection) [7].

Transmitter	Frequency	Maximum Output Power (kW)	Typical Antenna Gain (dBi)	Radius of the Vulnerable Zone in the Main Beam (m)
Marine VHF fixed	156 MHz to 174 MHz	0.025	3	1.1

Both initial assessment on the general criteria and initial assessment specific to the offshore plant were performed on the transmission. The results on land were more conservative, with the vulnerable area of 6 m, compared to the 1.1 m of the offshore plant.

3.2. Full Assessment

Full assessment should be carried out in the initial assessment when a potential hazard appears and proceeds as follows:

- Collect all relevant information on transmissions and the plant
- Calculate the effective field strength taking into account effects of modulation
- Calculate the extractable power or energy from the adventitious antenna
- Compare the extractable power or energy from the adventitious antenna with the threshold values

Because most transmissions are modulated, the modulation must be made with clearance, and the calculated field strengths use the modulation factor to obtain the effective field strength.

The transmission is subjected to full assessment up to 6 m in the vulnerable area where the initial assessment result is more conservative. Effective field strengths are classified into three categories based on 10.4.3 [7] of the BS guide. The first one is horizontal polarization with a frequency of less than 30 MHz. The second one is vertical polarization with frequency below 30 MHz. The third one is when the frequency exceeds 30 MHz. The effective field strength is calculated by selecting one of three equations for each transmitter condition [12]. Since the frequency of the transmission is more than 30 MHz at 163 MHz, the effective field strength is calculated as follows:

$$E = \frac{0.173 mF \sqrt{(PG)}}{d} \tag{1}$$

In this case, the modulation factor (m) of the VHF is 1.0 as a frequency of phase modulation (FM), and the horizontal radiation pattern (F) is 1. The antenna gain (G) is measured by the directional power (dB). If the gain is expressed in decibels, it is calculated by the following equation of BS guide [7]:

$$G = 10^{0.1g} \quad (2)$$

The calculated effective field strengths are quantified in Table 4.

Table 4. Effective field strength for single transmissions.

No.	m	F	P (kW)	g (dBi)	G	d (km)	E (V/m)
1	1	1	0.1	3	1.995	0.001	77.3
2	1	1	0.1	3	1.995	0.002	38.6
3	1	1	0.1	3	1.995	0.003	25.8
4	1	1	0.1	3	1.995	0.0035	22.1
5	1	1	0.1	3	1.995	0.004	19.3
6	1	1	0.1	3	1.995	0.005	15.5
7	1	1	0.1	3	1.995	0.006	12.9

The extractable power is calculated considering the inner circumference of the loop-type structure when the frequency is 30 MHz or less. In contrast, for frequencies above 30 MHz, without considering the perimeter of the structure, the equation given below is calculated, taking into account only the effective field strength and frequency.

$$P_{max} = \frac{124E^2}{f^2 + 3030} \quad (3)$$

The calculated extractable power is shown in Table 5.

Table 5. Extractable power for single transmissions.

No.	f (MHz)	d (km)	E (V/m)	P_{max} (W)
1	163	0.001	77.3	25.02
2	163	0.002	38.6	6.25
3	163	0.003	25.8	2.78
4	163	0.0035	22.1	2.04
5	163	0.004	19.3	1.56
6	163	0.005	15.5	1.00
7	163	0.006	12.9	0.69

The calculated power is compared to the threshold value specified in the BS guide. Table 6 shows the threshold power and thermal initiation time criterion from the BS guide. In this paper, the risk of ignition was evaluated using threshold power. If the P_{max} calculated for each gas group is less than Pth on the BS, the evaluation is no longer carried out because there is no risk of RF ignition. If P_{max} is greater than or equal to Pth, there is a potential risk of high-frequency ignition.

Table 6. Radio frequency power thresholds [7].

Gas Group	Threshold Power, Pth (W)	Thermal Initiation Time (μs)	Representative Gas
I	6 for long narrow structures, e.g., cranes; 8 for all other structures	200	Methane
IIA	6	100	Propane
IIB	3.5	80	Ethylene
IIC	2	20	Hydrogen

The gas type is assumed to be the IIC group, considering the hydrogen as a high-risk gas located in the offshore plant, and the results are compared. As a result, of evaluating risk of ignition depending on distance, there is a risk of high-frequency ignition because P_{max} is equal to or larger than Pth from the transmission to 3.5 m. In contrast, from 4 m to 6 m, P_{max} is smaller than Pth, so there is no danger of high-frequency ignition.

The full assessment result and the initial assessment result at 30 MHz or more were compared with each other. In the initial evaluation using the general criteria, it was judged that there was a risk of ignition within 6 m of the vulnerable zone, and there was a danger from the full assessment to 3.5 m after the initial assessment. However, in the initial assessment, which applied specific criteria for offshore plants, it was judged that there was a risk of ignition within 1.1 m of the vulnerable zone. This analysis shows that the BS guide on general criteria including land and sea produces more conservative results, and that marine plant-specific criteria are limited.

3.3. Multiple Transmission Assessment

In the case of multiple transmissions, the effective field strength and P_{max} for each transmission are calculated according to frequency. If the sum of all the values of P_{max} is less than Pth, there is no risk of RF ignition, and the assessment can be stopped. However, if the sum of these values of P_{max} is larger than Pth, then the off-resonance effects should be considered by calculation of the modulus match power, P_{mm}. P_{mm} can be obtained using the following equation, where f_r is the resonant frequency of structure, f_t is the transmission frequency, and Q_k is the circuit factor:

$$\frac{P_{mm}}{P_{max}} = \frac{2}{1+n} \quad (4)$$

$$n = \frac{Q_k \left[1 + \left\{Q_k - \left(Q_k + \frac{1}{Q_k}\right)\left(\frac{f_r}{f_t}\right)^2\right\}^2\right]^{1/2}}{(1+Q_k^2)^{1/2}\left(\frac{f_r}{f_t}\right)^2} \quad (5)$$

Q_k (quality factor or circuit factor) [13–15] at resonance is the quality of the frequency selection characteristic and is calculated using the equation below. The resonance frequency divided by the 3 dB bandwidth on both sides is Q_k, which means that the band is wide when the value is low and narrow when it is high.

$$Q_k = \frac{\text{resonance frequency}}{\text{3dB Bandwidth}} \quad (6)$$

f_t is the frequency of the transmission used in the assessment, 163 MHz, and f_r and Q_k are obtained through calculation. f_r is the resonance frequency of the structure and is determined to be the most conservative value at which the ratio of f_t:f_r is 1.0 with reference to Figure 2. f_r is determined to be 163 MHz since f_t is 163 MHz. Figure 2 is taken from the BS guide, and the Q_k value is 5 as an example.

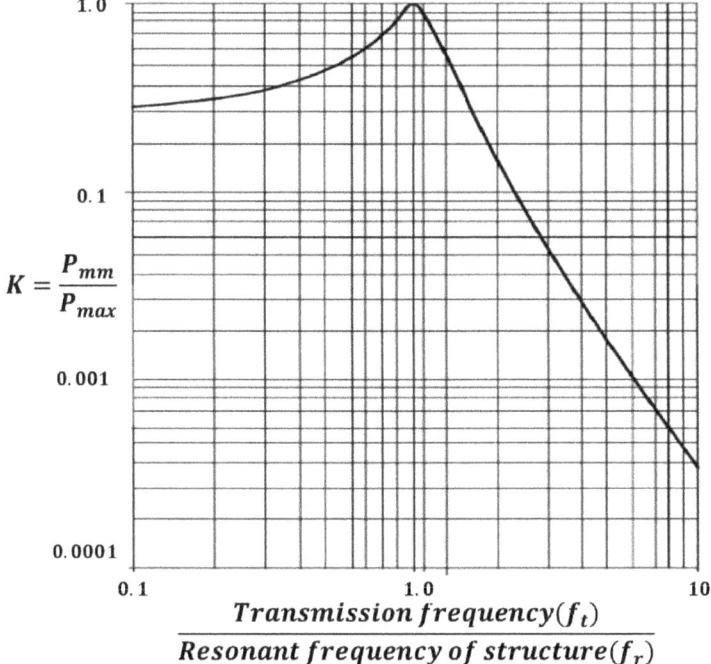

Figure 2. Modulus match powers.

As a result, if f_t and f_r are the most conservative values, n is 1 regardless of Q_k, according to the above Equation (4). Therefore, the value of P_{mm} is equal to the value of P_{max}. That is, the power of the multiplex transmission is the sum of P_{mm} of the transmitters as shown in Table 7. Comparing the threshold power and the sum of P_{mm} in Table 6, it is determined that there is a danger of high-frequency ignition because the sum of P_{mm} is equal to or greater than the threshold power up to a distance of 5 m. However, when the distance is more than 5 m, it is assumed that there is no danger of high-frequency ignition because P_{mm} is less than the critical power.

Table 7. Extractable power for multiple transmissions.

No.	f (MHz)	d (km)	P_{mm} (W)	Sum of P_{mm} (W)
1	163	0.001	25.02	50.03
2	163	0.002	6.25	12.51
3	163	0.003	2.78	5.56
4	163	0.0035	2.04	4.08
5	163	0.004	1.56	3.13
6	163	0.005	1.00	2.00
7	163	0.006	0.69	1.39

Therefore, the on-site test should be performed at the point, where it is judged if there is a risk of high-frequency ignition because the power value that can be extracted from the threshold power is greater than or equal to the threshold value. The on-site test is used to measure the actual extractable power and compare it with the threshold power of the gas group. if the results of on-site tests indicate a risk, we can investigate measures such as plant design changes, plant movement, transmission movement, and transmission power reduction. Methods include bonding, insulation, reducing the effectiveness of the structure as a receiving antenna, and de-tuning.

4. Numerical Analysis

To compare with the results based on the BS guide, the effective field strength and power are calculated by Ansys HFSS. Numerical analysis derives the effective field strength and power at a specific distance from the transmission without considering the structure to simplify the calculation. Numerical analysis was performed to measure the effective field strength up to 5 m because of the risk assessment of code analysis. Numerical analysis was performed for a single transmission and multiple transmissions. The actual shape of the dipole antenna, which is the basic transmission, and the modeled shape are shown in Figure 3. The detailed specifications are shown in Table 1.

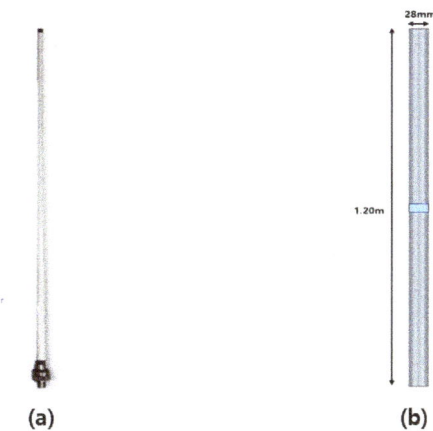

Figure 3. (a) Actual dipole antenna; (b) model for the dipole antenna.

In the numerical analysis, the effective field strength (E) of 1 m to 5 m from the transmission to the X-axis is measured from 0 to 360 degrees in 5-degree intervals according to phase and power (P). The effective field strength was calculated by assuming root mean square (RMS), which is an average value index widely used in antenna analysis. The RMS equation is defined as follows:

$$E_{rms} = \sqrt{\frac{E_1^2 + E_2^2 + \cdots + E_n^2}{n}} \tag{7}$$

The power passing through the surface of a specific location was calculated using the Ansys HFSS Fields Calculator. The defined power equation is as follows. Here, S is the surface used to calculate the force, and $\vec{n}$ is the normal vector for S.

$$W = \int_S Re(\vec{P}) \cdot \vec{n} \, dS \tag{8}$$

The Fields Calculator [16] referenced HFSS's Fields Calculator Cookbook's Fields Calculator Recipes.

4.1. Analysis for a Single Transmission

For a single transmission, the effective field strength (E) and power (P) were calculated from 1 m to 5 m from the transmission on the X-axis. The modeling of the transmission is shown in Figure 4, and the main specifications are as follows. Since the boundary conditions cannot be infinite in the analysis space, the boundaries are selected to be ±5.2 m on the X-, Y-, and Z-axes, and radiation conditions are given to the outermost surfaces. The port is a lumped condition, and the impedance is fed to the default value of 50 Ω. The frequency of the transmission is 163 MHz, the power is 100 W, the antenna length is 1200 mm, and the antenna thickness is 28 mm. The E field had an omni-directional shape, as shown in Figure 5. Table 8 shows the effective field strength and power measured by Ansys HFSS.

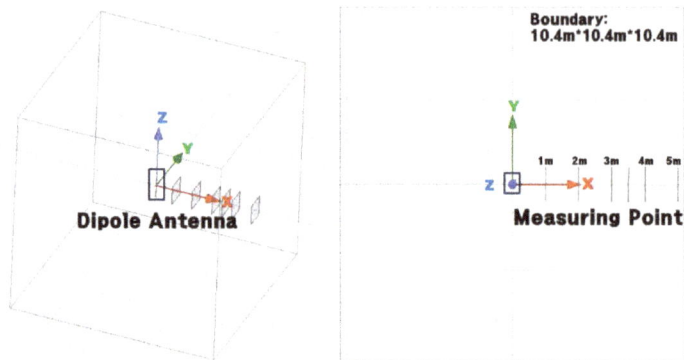

Figure 4. Modeling and measurement point for a single transmission.

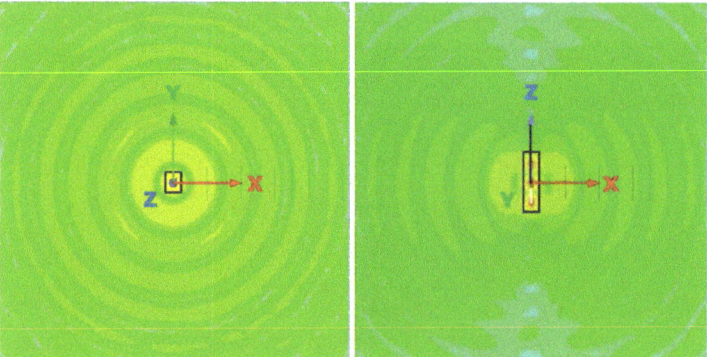

Figure 5. E field for a single transmission.

Table 8. Results for a single transmission.

No.	F (MHz)	P (kW)	d (km)	E (V/m)	P_{max} (W)
1	163	0.1	0.001	39.3	3.98
2	163	0.1	0.002	32.0	1.36
3	163	0.1	0.003	26.3	0.64
4	163	0.1	0.0035	23.8	0.48
5	163	0.1	0.004	22.3	0.37
6	163	0.1	0.005	17.6	0.25

4.2. Analysis of Multiple Transmissions

In the case of multiple transmissions, modeling is done as shown in Figure 6, and the condition is the same as a single transmission. The E field is shown in Figure 7. Unlike a single transmission, two antennas transmit to each other, producing interference. Both transmissions emit the same radiation and are located 4.5 m from each other. The effective field strength (E) and power (P) were calculated from the axis between the two transmissions, from 1 m to 5 m, on the same X-axis as the single antenna. The results of Ansys HFSS are shown in Table 9.

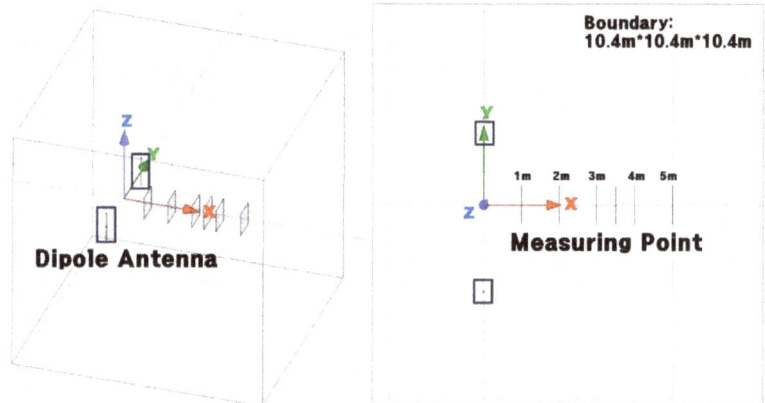

Figure 6. Modeling and measurement point for multiple transmissions.

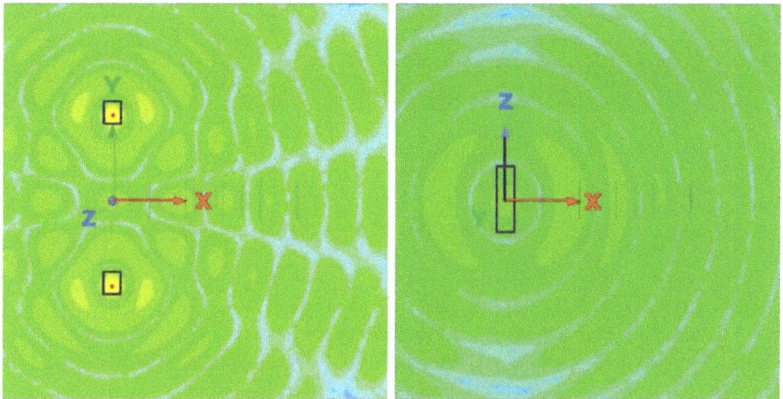

Figure 7. E field for multiple transmissions.

Table 9. Results for multiple transmissions.

No.	F (MHz)	P (kW)	d (km)	E (V/m)	P_{max} (W)
1	163	0.1	0.001	41.5	0.91
2	163	0.1	0.002	23.4	1.14
3	163	0.1	0.003	15.4	1.04
4	163	0.1	0.0035	13.9	0.97
5	163	0.1	0.004	11.7	0.86
6	163	0.1	0.005	10.0	0.66

5. Comparison and Analysis

In this paper, risk assessment was performed based on the BS guide and numerical analysis using Ansys HFSS to improve the accuracy of the results.

Table 10 compares the effective field strength and power for a single transmission. The code analysis calculates the effective field strength and power according to the entire procedure of the BS guide. Numerical analysis is performed by measuring the Ansys HFSS results. As a result, the effective field strength and power for a single transmission are calculated to be more conservative in the results based on the BS guide. Figure 8 shows comparison between P of BS guide and P of HFSS for a single transmission.

Table 10. Comparison between BS guide and HFSS for a single transmission.

No.	d (km)	E of the BS Guide (V/m)	E of HFSS (V/m)	P of the BS Guide (W)	P of HFSS (W)
1	0.001	77.3	39.3	25.02	3.98
2	0.002	38.6	32.0	6.25	1.36
3	0.003	25.8	26.3	2.78	0.64
4	0.0035	22.1	23.8	2.04	0.48
5	0.004	19.3	22.3	1.56	0.37
6	0.005	15.5	17.6	1.00	0.25

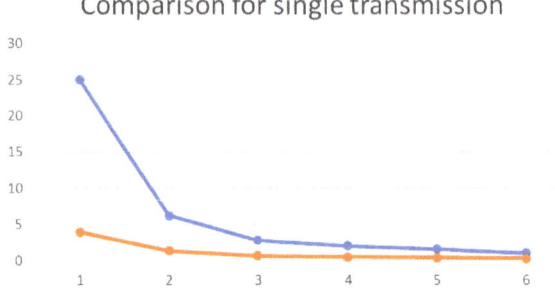

Figure 8. Comparison between P of BS guide and P of HFSS for a single transmission.

Table 11 compares the power for multiple transmissions. In the multiple transmission environment, it is possible to calculate the effective field strength and power in the numerical analysis. However, since the BS guide cannot calculate the effective field strength, only power can be calculated. Therefore, only power values are compared in the multiple transmission environment. The results are calculated at the same point as for the single transmission. The power for multiple transmissions is calculated to be more conservative for the results based on the BS guide. Figure 9 shows comparison between the power of BS guide and that of HFSS for multiple transmission. The BS guide calculates the effective field strength by using the distance difference, the antenna gain, and the antenna power, and simply calculates the power considering the antenna frequency. On the other hand, the Ansys HFSS calculates the intensity of the electric power by calculating the 3D electromagnetic field formula. Therefore, it is expected that the accuracy of risk assessment based on the BS guide can be improved by using the proposed numerical analysis.

Table 11. Comparison between BS guide and HFSS for multiple transmissions.

No.	d (km)	P of the BS Guide (W)	P of HFSS (W)
1	0.001	50.03	0.91
2	0.002	12.51	1.14
3	0.003	5.56	1.04
4	0.0035	4.08	0.97
5	0.004	3.13	0.86
6	0.005	2.00	0.66

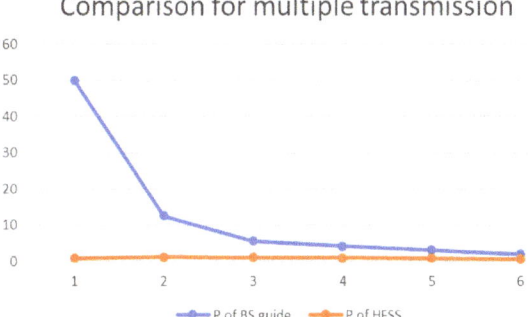

Figure 9. Comparison between the power of BS guide and that of HFSS for multiple transmission.

6. Conclusions

This paper proposed a numerical analysis method to improve the accuracy of risk assessment based on the BS guide. In this paper, risk assessment was performed based on BS PD CLC/TR 50427:2004 and IEC 6007 for a dipole antenna with an inner circumference of less than 85 m and a frequency of 30 MHz or more, which are commonly used in offshore plants. In addition, the proposed numerical analysis was performed using Ansys HFSS. Initial assessment and full assessment were performed according to the assessment procedure of the BS guide for the dipole antenna. Then, to simplify the calculation, the proposed numerical analysis was performed under conditions that did not consider the structure, and the effective field strength and power were derived. Comparing the effective field strength and power of a conventional code analysis and the proposed numerical analysis, the results of code analysis were more conservatively calculated. Experimental results showed that the proposed method using numerical analysis outperforms the code analysis in accuracy of the evaluation. n our observation, code analysis based on risk assessment was ineffective for practical site because it performed with the most general conditions without any consideration of actual environment such as structure. Therefore, it is possible to improve the accuracy of the result by using the proposed numerical analysis method.

Author Contributions: Conceptualization, K.I., Y.C. and D.S.; Methodology, J.K., K.I. and D.P.; Software, K.I., Y.C. and D.S.; Validation, K.I., Y.C., D.S., D.P. and J.K.; Formal Analysis, K.I. and Y.C.; Investigation, K.I., Y.C. and D.S.; Resources, K.I., Y.C., D.S., D.P. and J.K.; Data Curation, K.I., D.S. and Y.C.; Writing—Original Draft Preparation, Y.C.; Writing—Review and Editing, K.I. and Y.C.; Visualization, Y.C. and D.S.; Supervision, J.K., K.I. and D.P.; Project Administration, J.K., K.I. and D.P.; Funding Acquisition, J.K. and D.P.

Funding: This work was supported by Research Funds of Hyundai Heavy Industries. It was also supported by the Korea Institute of Energy Technology Evaluation and Planning (KETEP) and the Ministry of Trade, Industry, & Energy (MOTIE) of the Republic of Korea (No. 20172510102130).

Conflicts of Interest: The authors declare no conflict of interest.

References

1. Choi, S.W.; Kwon, H.M. Characteristics of induced voltage in loop structures from high-frequency radiation antenna. *J. Korean Soc. Saf.* **2014**, *29*, 49–54. [CrossRef]
2. Eckhoff, R.K.; Thomassen, O. Possible sources of ignition of potential explosive gas atmospheres on offshore process installations. *J. Loss Prev. Process Ind.* **1994**, *7*, 281–294. [CrossRef]
3. Bradby, I.R. *Practical Experience in Radio Frequency Induced Ignition Risk Assessment for COMAH/DSEAR Compliance*; ABB Engineering Services: Billingham, UK, 2008.

4. BS 6656:2002. Assessment of Inadvertent Ignition of Flammable Atmospheres by Radio-Frequency Radiation—Guide. 2002. Available online: https://landingpage.bsigroup.com/LandingPage/Undated?UPI=000000000030108909 (accessed on 27 November 2018).
5. Rajkumar, V.; Bhattacharjee, P. Risk assessment of RF radiation ignition hazard. In Proceedings of the International Conference on Reliability, Infocom Technologies and Optimization (ICRITO), Noida, India, 2–4 September 2015; pp. 1–3.
6. Wang, W.B.; Jiang, H.L.; Zhang, Y.P. Analysis of radio frequency risks in flammable and explosive environments. *Adv. Mater. Res.* **2015**, *1092–1093*, 717–721. [CrossRef]
7. BS PD CLC/TR 50427:2004. Assessment of Inadvertent Ignition of Flammable Atmospheres by Radio-Frequency Radiation—Guide. 2004. Available online: https://www.thenbs.com/PublicationIndex/documents/details?Pub=BSI&DocId=304016 (accessed on 27 November 2018).
8. IEC 60079-0. Explosive atmospheres—Part 0: Equipment—General requirements. 2011. Available online: https://webstore.iec.ch/publication/620 (accessed on 27 November 2018).
9. IEC 60079-10-1. Explosive atmospheres—Part 10-1: Classification of areas—Explosive gas atmospheres. 2008. Available online: https://webstore.iec.ch/publication/622 (accessed on 27 November 2018).
10. IEC 60079-10-2. Explosive atmospheres—Part 10-2: Classification of areas—Combustible dust atmospheres. 2009. Available online: https://webstore.iec.ch/publication/12903 (accessed on 27 November 2018).
11. *Product Brochure Marine Antennas*; AEP Marine Parts: Alblasserdam, Netherlands, 2016.
12. Shinn, D.H. Avoidance of radiation hazards from microwave antennas. *Marconi. Rev.* **1976**, *39*, 61–80.
13. Knight, P. *Radio Frequency Ignition Hazards: The Power Available from Non-Resonant Structures*; BBC Research Department Report No. RD 1982/3; British Broadcasting Corporation: London, UK, 1982.
14. Knight, P. *Radio Frequency Ignition Hazards: The Choice of Reference Antenna for Available-Power Calculations*; BBC Research Department Report No. RD 1982/16; British Broadcasting Corporation: London, UK, 1982.
15. Widginton, D.W. *Radio Frequency Ignition hazards, intrinsic safety*; Health and Safety Executive Research and Laboratory Services Division: London, UK, 1979.
16. ANSYS. *HFSS Fields Calculator Cookbook*; ANSYS Electromagnetics Suite 19.0; ANSYS: Canonsburg, PA, USA, 2017; pp. 29–30.

© 2018 by the authors. Licensee MDPI, Basel, Switzerland. This article is an open access article distributed under the terms and conditions of the Creative Commons Attribution (CC BY) license (http://creativecommons.org/licenses/by/4.0/).

Article

A Complex Lie-Symmetry Approach to Calculate First Integrals and Their Numerical Preservation

Wajeeha Irshad [1], Yousaf Habib [1,2,*] and Muhammad Umar Farooq [3]

[1] School of Natural Sciences, National University of Sciences and Technology, Sector H-12, Islamabad 44000, Pakistan; wajeehairshad97@gmail.com
[2] Department of Mathematics, COMSATS University Islamabad, Lahore Campus, Lahore 54000, Pakistan
[3] Department of BS and H, College of E and ME, National University of Sciences and Technology, Peshawar Road, Rawalpindi 46000, Pakistan; m_ufarooq@yahoo.com
* Correspondence: yhabib@gmail.com

Received: 26 September 2018; Accepted: 31 October 2018; Published: 24 December 2018

Abstract: We calculated Noether-like operators and first integrals of a scalar second-order ordinary differential equation using the complex Lie-symmetry method. We numerically integrated the equations using a symplectic Runge–Kutta method. It was seen that these structure-preserving numerical methods provide qualitatively correct numerical results, and good preservation of first integrals is obtained.

Keywords: Hamiltonian system; complex Lagrangian; Noether symmetries; first integrals; symplectic Runge–Kutta methods

MSC: 34C14; 37K05; 70G65; 65L05; 65L06

---

1. Introduction

Marius Sophus Lie proposed a symmetry-based method for the analytical solution of differential equations using groups of continuous transformations known as Lie groups [1–4]. Amalie Emmy Noether later presented her remarkable theorem that relates variational symmetries with conservation laws or first integrals in Reference [5]. In the literature, different methods are available to calculate first integrals of ordinary differential equations (ODEs), including the direct method, the characteristic or multiplier method, the Noether approach, and the partial Noether approach [6–9]. In this paper, we used the classical Noether approach to calculate the first integrals of a harmonic oscillator. We then applied the complex symmetry method in the restricted domain to find the first integrals of a system of harmonic oscillators by considering the Lagrangian in the complex variable domain [10–12].

Concerning the numerical solutions of ODEs with quadratic first integrals, it is well known that symplectic numerical methods are a suitable candidate [13]. These methods are a subclass of geometric integrators that preserve the geometric properties of the exact flow of ODEs. One class of symplectic methods with optimal order are the Gauss–Legendre Runge–Kutta methods. They are one-step numerical methods for ODEs and preserve all linear and quadratic first integrals of a dynamic system [14]. If we intend to preserve cubic or higher-order first integrals, we do not have a general numerical scheme for such a purpose, but we can design a numerical method that has this as its specific goal, for example, with the splitting and discrete-gradient methods [14]. In this paper, we present a way of constructing symplectic Runge–Kutta methods. We then take fourth-order Gauss–Legendre Runge–Kutta methods for the numerical integration of ODEs and report good preservation of first integrals by the numerical solution.

2. Symmetries and First Integrals

Consider a second-order ordinary differential equation,

$$\frac{d^2y}{dt} = f(t, y', y), \qquad (1)$$

which admits a Lagrangian L satisfying the Euler–Lagrange equation,

$$\frac{d}{dt}\left(\frac{\partial L}{\partial y'}\right) - \frac{\partial L}{\partial y} = 0. \qquad (2)$$

To explain the invariance criteria for variational problems under a group of transformation, we consider the operator

$$X = \xi(t, y)\frac{\partial}{\partial t} + \eta(t, y)\frac{\partial}{\partial y}, \qquad (3)$$

where X is the Noether symmetry generator for the Lagrangian L with gauge function $B(t, y)$, provided the following condition holds,

$$X^{(1)}(L) + D_t(\xi)L - D_t(B) = 0, \qquad (4)$$

where $X^{(1)}$ is a first-order prolongation of X and D represents total derivative,

$$D_t = \frac{\partial}{\partial t} + y'\frac{\partial}{\partial y}. \qquad (5)$$

According to the Noether theorem, for each Noether symmetry of a Euler–Lagrange equation, there corresponds a function I

$$I = \xi L + (\eta - \xi y')\frac{\partial L}{\partial y'} - B(t, y), \qquad (6)$$

called the first integral or conserved quantity of Equation (1) with respect to symmetry generator X.

Complex Symmetry Analysis

We first discuss some important results related to complex Noether symmetries, complex Lagrangian, and the Noether theorem in the restricted complex domain. We use them to determine first integrals of second-order restricted complex ODEs [15]. We then present expressions for Euler–Lagrange-like equations, conditions for Noether-like operators, and expressions for first integrals corresponding to these operators. For more details, see Reference [10] and references therein.

Consider a system of two second-order ordinary differential equations of the form

$$\begin{aligned}\frac{d^2f}{dt} &= w_1(t, g, f, g', f'),\\ \frac{d^2g}{dt} &= w_2(t, g, f, g', f').\end{aligned} \qquad (7)$$

Suppose we have a transformation $y(t) = f + ig$ and $w = w_1 + iw_2$, which converts System (7) to a second-order restricted complex ODE,

$$y'' = w(t, y, y'). \qquad (8)$$

Assume that Equation (8) admits a complex Lagrangian $L(t, f, g, f', g')$, i.e., $L = L_1 + iL_2$. Therefore, we have two Lagrangians, L_1 and L_2, for System (7) that satisfy Euler–Lagrange-like equations:

$$\frac{\partial L_1}{\partial f} + \frac{\partial L_2}{\partial g} - \frac{d}{dt}\left(\frac{\partial L_1}{\partial f'} + \frac{\partial L_1}{\partial g'}\right) = 0,$$
$$\frac{\partial L_2}{\partial f} - \frac{\partial L_1}{\partial g} - \frac{d}{dt}\left(\frac{\partial L_2}{\partial f'} - \frac{\partial L_1}{\partial g'}\right) = 0. \tag{9}$$

The operators

$$X_1 = \varsigma_1 \frac{\partial}{\partial t} + \chi_1 \frac{\partial}{\partial f} + \chi_2 \frac{\partial}{\partial g},$$
$$X_2 = \varsigma_2 \frac{\partial}{\partial t} + \chi_2 \frac{\partial}{\partial f} - \chi_1 \frac{\partial}{\partial g}. \tag{10}$$

are called Noether-like operators for Lagrangians L_1 and L_2 such that:

$$-X_2^{(1)}(L_2) + X_1^{(1)}(L_1) + (D_t\varsigma_1)L_1 - (D_t\varsigma_2)L_2 = D_t A_1,$$
$$X_2^{(1)}(L_1) + X_1^{(1)}(L_2) + (D_t\varsigma_1)L_2 + (D_t\varsigma_2)L_1 = D_t A_2, \tag{11}$$

where A_1 and A_2 are suitable gauge functions. The two first integrals corresponding to Noether-like operators X_1 and X_2 can be found as:

$$I_1 = -A_1 + \varsigma_1 L_1 + \partial_{f'}L_1(\chi_1 - \varsigma_2 L_2 - \varsigma_1 f' - \varsigma_2 g') - \partial_{f'}L_2(\chi_2 - \varsigma_2 f' - \varsigma_1 g'),$$
$$I_2 = -A_2 + \varsigma_1 L_2 + \partial_{f'}L_2(\chi_1 + \varsigma_2 L_1 - \varsigma_1 f' - \varsigma_2 g') + \partial_{f'}L_1(\chi_2 - \varsigma_2 f' - \varsigma_1 g'). \tag{12}$$

3. Runge–Kutta Methods

Runge–Kutta methods [16] are one-step numerical methods for the approximate solution of IVPs:

$$y'(t) = f(y(t)), \quad y(t_0) = y_0, \quad y(t) \in \mathbb{R}^n. \tag{13}$$

These methods provide approximation $y_n = y(t_n)$ of the exact solution $y(t)$ at time $t_n = nh$, where $n = 0, 1, \cdots$ and h corresponds to the stepsize. The generalized form of an s-stage Runge–Kutta method is

$$Y_k = y_{n-1} + \sum_{i=1}^{s} a_{ki} h f(Y_i), \quad k = 1, \cdots, s, \tag{14}$$
$$y_n = y_{n-1} + \sum_{l=1}^{s} b_l h f(Y_l),$$

with b_i representing the weights and c_i, the nodes at which stages Y_k are evaluated. A Runge–Kutta method can be represented by a Butcher tableau:

c_1	a_{11}	$\cdots$	a_{1n}
$\vdots$	$\vdots$	$\ddots$	$\vdots$
c_n	a_{n1}	$\cdots$	a_{nn}
	b_1	$\cdots$	b_n

For explicit Runge–Kutta methods, we have $a_{ki} = 0$ for $k \leq i$,; otherwise, they are implicit.

3.1. Symplectic Runge–Kutta Methods

If Equation (13) has a quadratic first integral

$$I(y) = \langle y, Sy \rangle = y^T Sy,$$

where S is a symmetric square matrix, then we have

$$\langle y, f(y) \rangle = y^T S f(y) = 0.$$

We want to determine numerical solutions y_n such that first integral $I(y)$ is preserved numerically, i.e.,

$$\langle y_n, Sy_n \rangle = \langle y_{n-1}, Sy_{n-1} \rangle \quad n = 0, 1, \ldots.$$

It has been shown in References [17–19] that only symplectic Runge–Kutta methods preserve quadratic first integrals while numerically integrating System (13). Moreover, in this paper we are only considering implicit Runge–Kutta methods to check the numerical preservation of first integrals because explicit methods cannot be symplectic [20]. A Runge–Kutta method is symplectic if its coefficients satisfy the following condition [18,19,21]:

$$b_i a_{ij} + b_j a_{ji} - b_i b_j = 0 \ \forall \ j, i = 1, \ldots, s, \tag{15}$$

which can be derived as follows.

Firstly, apply the Runge–Kutta method (14) to solve the IVP (13). The stage values are

$$Y_i = y_{n-1} + \sum_j h a_{ij} f(Y_j).$$

Since

$$\langle Y_i, Sf(Y_i) \rangle = 0,$$
$$\Rightarrow \langle y_{n-1}, Sf(Y_i) \rangle + \sum_j h a_{ij} \langle f(Y_j), Sf(Y_i) \rangle = 0. \tag{16}$$

Moreover, for the output values, we have

$$y_n = y_{n-1} + \sum_{i=1}^{s} b_i h f(Y_i).$$

Thus,

$$\langle y_n, Sy_n \rangle = \langle y_{n-1}, Sy_{n-1} \rangle + h \sum_i b_i \langle y_{n-1}, Sf(Y_i) \rangle$$
$$+ h \sum_j b_j \langle f(Y_j), Sy_{n-1} \rangle + h^2 \sum_{i,j} b_j b_i \langle f(Y_i), Sf(Y_j) \rangle. \tag{17}$$

Evidently from Systems (16) and (17), we have

$$\langle y_n, Sy_n \rangle = \langle y_{n-1}, Sy_{n-1} \rangle,$$

provided that

$$b_j a_{ji} + b_i a_{ij} - b_i b_j = 0. \tag{18}$$

3.2. Construction of Symplectic RK Methods

Although there exist several techniques to construct symplectic RK methods in the literature [14,22], here we constructed symplectic Runge–Kutta methods with the help of a Vandermonde transformation. This was first discussed in reference [23].

A Vandermonde matrix is given as

$$V = \begin{bmatrix} 1 & c_1 & \cdots & c_1^{n-1} \\ \vdots & \vdots & \ddots & \vdots \\ 1 & c_n & \cdots & c_n^{n-1} \end{bmatrix} = c_i^{j-1}.$$

Pre- and postmultiply, Vandermonde matrix V with symplectic condition (15) as

$$c_i^{k-1}(b_j a_{ji} + b_i a_{ij} - b_j b_i)c_j^{l-1} = 0, \quad \forall \, l, k, j, i = 1, 2, \ldots, s. \tag{19}$$

To construct methods with two stages ($s = 2$), we consider
For $l, k = 1$,

$$\sum_{i,j}(b_j a_{ji} + b_i a_{ij} - b_j b_i) = 0. \tag{20}$$

For $l = 1$ and $k = 2$,

$$\sum_{i,j}(b_j c_j a_{ji} + b_i a_{ij} c_j - b_j c_j b_i) = 0. \tag{21}$$

For $l = 2$ and $k = 1$,

$$\sum_{i,j}(b_i a_{ij} c_j + b_j c_j a_{ji} - b_i b_j c_j) = 0. \tag{22}$$

For $l, k = 2$,

$$\sum_{i,j}(b_i c_i a_{ij} c_j + b_j c_j a_{ji} c_i - b_i c_i b_j c_j) = 0. \tag{23}$$

The following order two conditions must be satisfied.

$$\sum_j b_j = 1, \quad \sum_j b_j c_j = \frac{1}{2}. \tag{24}$$

Using Equation (24) in Equations (20)–(23), we have

$$\sum_i b_i c_i = \tfrac{1}{2},$$

$$\sum_{i,j=1}^{2}(b_i a_{ij} c_j + b_j c_j a_{ji}) = \tfrac{1}{2},$$

$$\sum_{i,j=1}^{2}(b_i c_i a_{ij} + b_j a_{ji} c_i) = \tfrac{1}{2},$$

$$\sum_{i,j=1}^{2} b_i c_i a_{ij} c_j = \tfrac{1}{8}.$$

If we take $b_i(c_i - c_1) = b_i c_i - b_i c_1$ and take summation of i from 1 to 2, we get

$$b_2 c_2 - b_2 c_1 = \frac{1}{2} - c_1,$$

$$b_2 = \frac{c_1 - \tfrac{1}{2}}{c_1 - c_2}.$$

Similarly,
$$b_1 = \frac{c_2 - \frac{1}{2}}{c_2 - c_1}.$$

If we take the relation
$$b_i(c_j - c_1)a_{ij}(c_i - c_1) = b_i c_i a_{ij} c_j - b_i c_i a_{ij} c_1 - b_i a_{ij} c_j c_1 + b_i a_{ij} c_1 c_1.$$

Thus, we get
$$a_{22} = \frac{\frac{1}{8} - \frac{c_1}{6} - \frac{c_1}{3} + \frac{c_1 c_1}{2}}{b_2(c_2 - c_1)^2}.$$

Similarly, we get
$$a_{11} = \frac{\frac{1}{8} - \frac{c_2}{6} - \frac{c_2}{3} + \frac{c_2 c_2}{2}}{b_1(c_1 - c_2)^2},$$

$$a_{21} = \frac{\frac{1}{8} - \frac{c_2}{3} - \frac{c_1}{6} + \frac{c_1 c_2}{2}}{b_2(c_2 - c_1)(c_1 - c_2)},$$

$$a_{12} = \frac{\frac{1}{8} - \frac{c_1}{3} - \frac{c_2}{6} + \frac{c_1 c_2}{2}}{b_1(c_2 - c_1)(c_1 - c_2)}.$$

Let us consider the shifted Legendre polynomials P_t^* on the interval $[0,1]$,

$$P_t^*(y) = \sum_{n=0}^{t} \frac{t!}{2t} \binom{t}{n} \binom{t+n}{n} (-1)^{t-n} y^n.$$

For Gauss methods, we choose abscissa c_i as zeros of P_t^* which have an order $2t$. For Radau methods, we choose either $c_1 = 0$ or $c_t = 1$, or both of them and then take for Radau I methods, the abscissa as the zeros of the polynomial $P_{t-1}^*(y) + P_t^*(y)$ of order $2t - 1$. Similarly, for Radau II methods, we take the abscissa as the zeros of the polynomial $P_t^*(y) - P_{t-1}^*(y)$ of order $2t - 1$. Moreover, for Lobatto III methods, we take the abscissa as the zeros of the polynomial $P_t^*(y) - P_{t-2}^*(y)$ of order $2t - 2$. Thus, we have the following symplectic methods:

Gauss, s = 2:

$$\begin{array}{c|cc}
\frac{1}{2} - \frac{\sqrt{3}}{6} & \frac{1}{4} & \frac{1}{4} - \frac{\sqrt{3}}{6} \\
\frac{1}{2} + \frac{\sqrt{3}}{6} & \frac{1}{4} + \frac{\sqrt{3}}{6} & \frac{1}{4} \\
\hline
 & \frac{1}{2} & \frac{1}{2}
\end{array}$$

Radau I, s = 2:

$$\begin{array}{c|cc}
0 & \frac{1}{8} & \frac{-1}{8} \\
\frac{2}{3} & \frac{7}{24} & \frac{3}{8} \\
\hline
 & \frac{1}{4} & \frac{3}{4}
\end{array}$$

Radau II, s=2:

$$\begin{array}{c|cc} \frac{1}{3} & \frac{3}{8} & \frac{-1}{24} \\ 1 & \frac{7}{8} & \frac{1}{8} \\ \hline & \frac{3}{4} & \frac{1}{4} \end{array}$$

Similarly, we can construct methods with more stages and a higher order.

4. Construction of First Integrals and Their Numerical Preservation

We construct the first integrals of a system of harmonic oscillators (both coupled and uncoupled) determined by the second-order ODE:

$$y'' = -k^2 y. \tag{25}$$

We take different values of k and y, as follows:

Case I: ($k^2 = 1$ and y is real)

When $k^2 = 1$ and $y(t)$ is real-valued, (25) becomes a one-dimensional harmonic oscillator equation:

$$y'' = -y, \tag{26}$$

that possesses the standard Lagrangian

$$L = \frac{y'^2}{2} - \frac{y^2}{2}. \tag{27}$$

Taking the Lagrangian and inserting it in System (4) yields the following determining system of equations:

$$-\eta y + \eta_t y' + (\eta_y - \frac{1}{2}\xi_t)y'^2 - \frac{1}{2}\xi_y y'^3 - \frac{1}{2}\xi_t y^2 - \frac{1}{2}\xi_y y^2 y' - B_t - y' B_y = 0. \tag{28}$$

Comparing different powers of y', we have a system of four partial differential equations whose solution gives rise to:

$$\begin{aligned} \xi(t,y) &= c_1 + c_2 \sin(2t) + c_3 \cos(2t), \\ \eta(t,y) &= \sin t\, c_4 + (\cos(2t)y\, c_2 - \sin(2t)y\, c_3) + \cos t\, c_5, \\ B(t,y) &= -(c_2 \sin(2t) + c_3 \cos(2t))y^2 + (c_4 \cos t - c_5 \sin t)y. \end{aligned} \tag{29}$$

We thus obtain the following 5-Noether symmetry generators:

$$\begin{aligned} X_1 &= \frac{\partial}{\partial t}, \\ X_2 &= \sin(2t)\frac{\partial}{\partial t} + y\cos(2t)\frac{\partial}{\partial y}, \\ X_3 &= \cos(2t)\frac{\partial}{\partial t} - y\sin(2t)\frac{\partial}{\partial y}, \\ X_4 &= \cos(t)\frac{\partial}{\partial y}, \\ X_5 &= \sin(t)\frac{\partial}{\partial y}. \end{aligned} \tag{30}$$

Using Symmetries (30) and Lagrangian (27) in Noether's Theorem (6), we obtain the following first integrals:

$$\begin{aligned} I_1 &= \frac{y'^2}{2} + \frac{y^2}{2}, \\ I_2 &= y \sin t + y' \cos t, \\ I_3 &= -y \cos t + y' \sin t, \\ I_4 &= -\frac{1}{2} y'^2 \cos 2t - y y' \sin 2t + \frac{1}{2} y^2 \cos 2t, \\ I_5 &= -\frac{1}{2} y'^2 \sin 2t + y y' \cos 2t + \frac{1}{2} y^2 \sin 2t. \end{aligned} \tag{31}$$

Among these five first integrals, only two are independent [8]. We numerically integrate system (26) using a fourth-order Gauss $s = 2$ symplectic Runge–Kutta method that we refer to from now on as Gauss2. We compare the results of the Gauss2 method with the famous symplectic Euler method [14], given as:

$$\begin{aligned} U_{n+1} &= U_n + h f(V_n), \\ V_{n+1} &= V_n - h g(U_{n+1}), \end{aligned}$$

for numerically integrating $U' = f(V)$ and $V' = g(U)$. We take stepsize $h = 0.01$, and $n = 10{,}000$ number of steps. By employing symplectic integrators, we expect the first integrals of the system to be preserved by the numerical schemes, and this is what we have achieved. We look at the deviation of numerically evaluated first integral $I(y_n)$ from the actual value of first integral $I(y_0)$. We calculate error by taking the difference of the first integral evaluated at initial value $I(y_0)$ with the value of the first integral evaluated at all subsequent numerically approximated values $I(y_n)$ given by the formula Error $= |I(y_n) - I(y_0)|$. Figures 1 and 2 represent the absolute error in integral I_2 using the Gauss2 and symplectic Euler method, respectively. Similarly, Figures 3 and 4 represent the absolute error in integral I_3 using the Gauss2 and symplectic Euler method, respectively. It is clear from the figures that the error is very small and bounded, depicting qualitatively correct numerical results. It is worth noting that the error of the Gauss2 method is much less compared to the error of the symplectic Euler method. The reason is that the Gauss2 method is fourth-order and more accurate compared to symplectic Euler method, which has order 1. Similar error behavior is obtained for other first integrals.

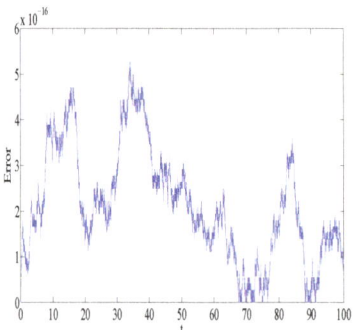

Figure 1. Error in integral I_2 using Gauss2.

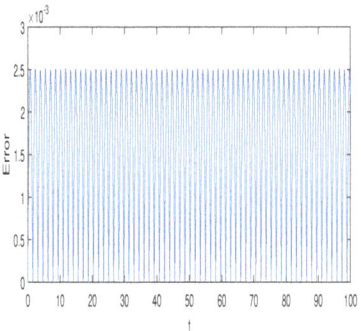

Figure 2. Error in integral I_2 using symplectic Euler.

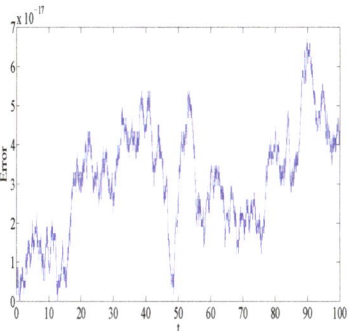

Figure 3. Error in integral I_3 using Gauss2.

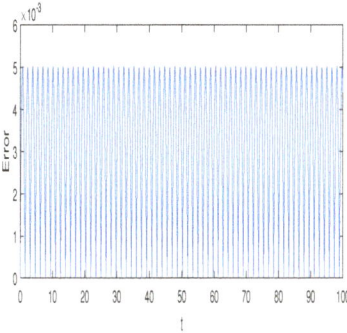

Figure 4. Error in integral I_3 using symplectic Euler.

Case II: ($k^2 = 1$ and y is complex)

When $k^2 = 1$ and $y(t)$ is a complex function $y = f + ig$ for f and g being real functions of t, we get:

$$f'' = -f, \qquad g'' = -g, \qquad (32)$$

which admits the following Lagrangians:

$$L_1 = \frac{1}{2}(-g'g' + f'f' - ff + gg),$$
$$L_2 = g'f' - fg. \qquad (33)$$

Using Lagrangians (33) in (11), we obtain 9-Noether-like operators

$$X_1 = \frac{\partial}{\partial t}, \quad X_2 = \sin t \frac{\partial}{\partial f}, \quad X_3 = \sin t \frac{\partial}{\partial g}, \quad X_4 = \cos t \frac{\partial}{\partial f}, \quad X_5 = \cos t \frac{\partial}{\partial g},$$

$$X_6 = \sin 2t \frac{\partial}{\partial t} + f \cos 2t \frac{\partial}{\partial f} + g \cos 2t \frac{\partial}{\partial g},$$

$$X_7 = g \cos 2t \frac{\partial}{\partial f} - f \cos 2t \frac{\partial}{\partial g}, \tag{34}$$

$$X_8 = \cos 2t \frac{\partial}{\partial t} - f \sin 2t \frac{\partial}{\partial f} - g \sin 2t \frac{\partial}{\partial g},$$

$$X_9 = -g \sin 2t \frac{\partial}{\partial f} + f \sin 2t \frac{\partial}{\partial g}.$$

Invoking Equation (12), we obtain the following invariants:

$$\begin{aligned}
I_{1,1} &= (f'f' - g'g' - ff + gg) \sin 2t - 2(ff' - gg') \cos 2t, \\
I_{1,2} &= 2(f'g' - fg) \sin 2t - 2(fg' + f'g) \cos 2t, \\
I_{2,1} &= (f'f' - g'g' - ff + gg) \cos 2t + 2(f'f - g'g) \sin 2t, \\
I_{2,2} &= 2(f'g' - fg) \cos 2t + 2(fg' + f'g) \sin 2t, \\
I_{3,1} &= -2f' \cos t - 2f \sin t, \\
I_{3,2} &= -2g' \cos t - 2g \sin t, \\
I_{4,1} &= -2f' \sin t + 2f \cos t, \\
I_{4,2} &= -2g' \sin t + 2g \cos t, \\
I_{5,1} &= f'f' - g'g' - f^2 + g^2, \\
I_{5,2} &= 2g'f' + 2gf.
\end{aligned} \tag{35}$$

associated with Noether-like operators (34). System of Equation (32) is integrated using the Gauss2 method with stepsize $h = 0.01$ and $n = 10{,}000$ number of steps. The absolute error in first integrals $I_{2,1}$, $I_{2,2}$, $I_{4,1}$, and $I_{4,2}$ is plotted in Figures 5–8, respectively. Similar error behavior is obtained for $I_{1,1}$, $I_{1,2}$, $I_{3,1}$, $I_{3,2}$, $I_{5,1}$, and $I_{5,2}$. We observe that the error does not grow out of bounds, which shows that the numerical method can mimic the true qualitative feature of the dynamical system.

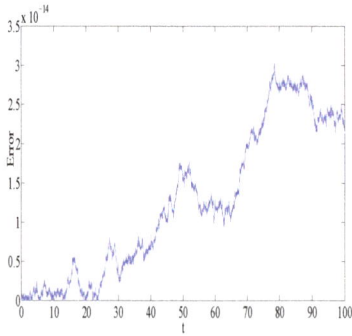

Figure 5. Error in integral $I_{2,1}$.

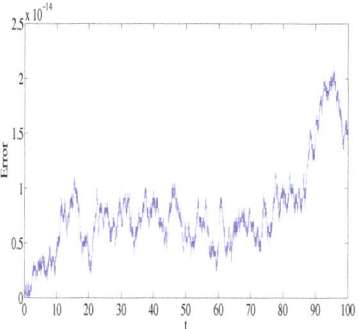

Figure 6. Error in integral $I_{2,2}$.

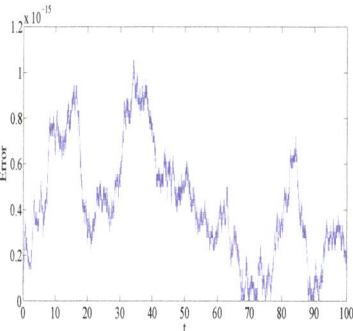

Figure 7. Error in integral $I_{4,1}$.

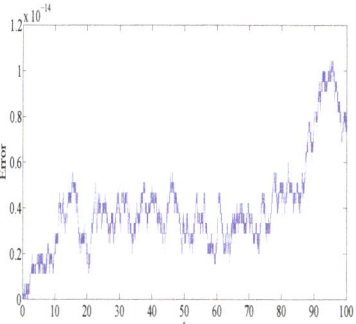

Figure 8. Error in integral $I_{4,2}$.

Case III: (k and y are complex)

When k and $y(t)$ are both complex, i.e., $k = \alpha_1 + i\alpha_2$ and $y = f + ig$ for f, g, α_1, and α_2 being real, the following coupled system of harmonic oscillators is obtained:

$$\begin{aligned} f'' &= -(\alpha_1^2 - \alpha_2^2)f + 2\alpha_1\alpha_2 g, \\ g'' &= -(\alpha_1^2 - \alpha_2^2)g - 2\alpha_1\alpha_2 f, \end{aligned} \tag{36}$$

which admits a pair of Lagrangians [11]:

$$L_1 = \frac{1}{2}f'^2 - \frac{1}{2}g'^2 - \frac{1}{2}(\alpha_1^2 - \alpha_2^2)(f^2 - g^2) + 2\alpha_1\alpha_2 fg,$$
$$L_2 = f'g' - \alpha_1\alpha_2(f^2 - g^2) - (\alpha_1^2 - \alpha_2^2)fg.$$

(37)

System (36) admits the following 9 Noether-like operators and first integrals:

$$X_1 = \frac{\partial}{\partial t},$$

$$X_2 = \sin(\alpha_1 t)\cosh(\alpha_2 t)\frac{\partial}{\partial f} + \cos(\alpha_1 t)\sinh(\alpha_2 t)\frac{\partial}{\partial g},$$

$$X_3 = \cos(\alpha_1 t)\sinh(\alpha_2 t)\frac{\partial}{\partial f} - \sin(\alpha_1 t)\cosh(\alpha_2 t)\frac{\partial}{\partial g},$$

$$X_4 = \cos(\alpha_1 t)\cosh(\alpha_2 t)\frac{\partial}{\partial f} - \sin(\alpha_1 t)\sinh(\alpha_2 t)\frac{\partial}{\partial g},$$

$$X_5 = -\sin(\alpha_1 t)\sinh(\alpha_2 t)\frac{\partial}{\partial f} - \cos(\alpha_1 t)\cosh(\alpha_2 t)\frac{\partial}{\partial g},$$

$$X_6 = \sin(2\alpha_1 t)\cosh(2\alpha_2 t)\frac{\partial}{\partial t} + \{(\alpha_1 f - \alpha_2 g)\cos(2\alpha_1 t)\cosh(2\alpha_2 t)$$
$$+ \{(\alpha_1 g + \alpha_2 f)\cos(2\alpha_1 t)\cosh(2\alpha_2 t) - (\alpha_1 f - \alpha_2 g)\sin(2\alpha_1 t)\sinh(2\alpha_2 t)\}\frac{\partial}{\partial g}$$
$$+ (\alpha_1 g + \alpha_2 f)\sin(2\alpha_1 t)\sinh(2\alpha_2 t)\}\frac{\partial}{\partial f},$$

$$X_7 = \cos(2\alpha_1 t)\sinh(2\alpha_2 t)\frac{\partial}{\partial t} + \{(\alpha_1 g + \alpha_2 f)\cos(2\alpha_1 t)\cosh(2\alpha_2 t)$$
$$- \{(\alpha_1 f - \alpha_2 g)\cos(2\alpha_1 t)\cosh(2\alpha_2 t) + (\alpha_1 g + \alpha_2 f)\sin(2\alpha_1 t)\sinh(2\alpha_2 t)\}\frac{\partial}{\partial g}$$
$$- (\alpha_1 f - \alpha_2 g)\sin(2\alpha_1 t)\sinh(2\alpha_2 t)\}\frac{\partial}{\partial f},$$

(38)

$$X_8 = \cos(2\alpha_1 t)\cosh(2\alpha_2 t)\frac{\partial}{\partial t} + \{(\alpha_1 f - \alpha_2 g)\sin(2\alpha_1 t)\cosh(2\alpha_2 t)$$
$$+ \{(\alpha_1 f - \alpha_2 g)\cos(2\alpha_1 t)\sinh(2\alpha_2 t) + (\alpha_1 g + \alpha_2 f)\sin(2\alpha_1 t)\cosh(2\alpha_2 t)\}\frac{\partial}{\partial g}$$
$$- (\alpha_1 g + \alpha_2 f)\cos(2\alpha_1 t)\sinh(2\alpha_2 t)\}\frac{\partial}{\partial f},$$

$$X_9 = -\sin(2\alpha_1 t)\sinh(2\alpha_2 t)\frac{\partial}{\partial t} + \{(\alpha_1 f - \alpha_2 g)\cos(2\alpha_1 t)\sinh(2\alpha_2 t)$$
$$- \{(\alpha_1 f - \alpha_2 g)\sin(2\alpha_1 t)\cosh(2\alpha_2 t) - (\alpha_1 g + \alpha_2 f)\cos(2\alpha_1 t)\sinh(2\alpha_2 t)\}\frac{\partial}{\partial g}$$
$$+ (\alpha_1 g + \alpha_2 f)\sin(2\alpha_1 t)\cosh(2\alpha_2 t)\}\frac{\partial}{\partial f}.$$

$$I_{1,1} = (\alpha_1^2 - \alpha_2^2)(f^2 - g^2) + f'^2 - 4\alpha_1\alpha_2 fg - g'^2,$$

$$I_{1,2} = 2(\alpha_1^2 - \alpha_2^2)fg + 2\alpha_1\alpha_2(f^2 - g^2) + 2f'g',$$

$$I_{2,1} = f'\sin(\alpha_1 t)\cosh(\alpha_2 t) - g'\cos(\alpha_1 t)\sinh(\alpha_2 t) - (\alpha_1 f - \alpha_2 g)\cos(\alpha_1 t)\cosh(\alpha_2 t)$$
$$- (\alpha_1 g + \alpha_2 f)\sin(\alpha_1 t)\sinh(\alpha_2 t),$$

$$I_{2,2} = g'\sin(\alpha_1 t)\cosh(\alpha_2 t) + f'\cos(\alpha_1 t)\sinh(\alpha_2 t) - (\alpha_1 g + \alpha_2 f)\cos(\alpha_1 t)\cosh(\alpha_2 t)$$
$$+ (\alpha_1 f - \alpha_2 g)\sin(\alpha_1 t)\sinh(\alpha_2 t),$$

$$I_{3,1} = f'\cos(\alpha_1 t)\cosh(\alpha_2 t) + g'\sin(\alpha_1 t)\sinh(\alpha_2 t) + (\alpha_1 f - \alpha_2 g)\sin(\alpha_1 t)\cosh(\alpha_2 t)$$
$$- (\alpha_1 g + \alpha_2 f)\cos(\alpha_1 t)\sinh(\alpha_2 t),$$

$$I_{3,2} = g'\cos(\alpha_1 t)\cosh(\alpha_2 t) - f'\sin(\alpha_1 t)\sinh(\alpha_2 t) + (\alpha_1 g + \alpha_2 f)\sin(\alpha_1 t)\cosh(\alpha_2 t)$$
$$+ (\alpha_1 f - \alpha_2 g)\cos(\alpha_1 t)\sinh(\alpha_2 t),$$

$$I_{4,1} = \frac{1}{2}[\{(\alpha_1^2 - \alpha_2^2)(-g^2 + f^2) - 4\alpha_1\alpha_2 gf - (-g'^2 + f'^2)\}\sin(2\alpha_1 t)\cosh(2\alpha_2 t)$$
$$- \{2\alpha_1\alpha_2 f^2 - 2\alpha_1\alpha_2 g^2 + 2(\alpha_1^2 - \alpha_2^2)gf - 2g'f'\}\cos(2\alpha_1 t)\sinh(2\alpha_2 t)]$$
$$+ \{\alpha_1(f'f - g'g) - \alpha_2(g'f + f'g)\}\cos(2\alpha_1 t)\cosh(2\alpha_2 t)$$
$$+ \{\alpha_1 g'f + \alpha_1 f'g + \alpha_2 f'f - \alpha_2 g'g\}\sin(2\alpha_1 t)\sinh(2\alpha_2 t)$$

$$I_{4,2} = \frac{1}{2}[\{(\alpha_1^2 - \alpha_2^2)(-g^2 + f^2) - 4\alpha_1\alpha_2 gf + (g'^2 - f'^2)\}\cos(2\alpha_1 t)\sinh(2\alpha_2 t)$$
$$+ \{2\alpha_1\alpha_2(f^2 - g^2) + 2fg(\alpha_1^2 - \alpha_2^2) - 2f'g'\}\sin(2\alpha_1 t)\cosh(2\alpha_2 t)]$$
$$+ [\{\alpha_1 f'g + \alpha_1 fg' + \alpha_2(f'f - gg')\}\cos(2\alpha_1 t)\cosh(2\alpha_2 t)]$$
$$- [\{\alpha_1 ff' - \alpha_1 g'g) - \alpha_2 g'f - \alpha_2 f'g)\}\sin(2\alpha_1 t)\sinh(2\alpha_2 t)]$$

$$I_{5,1} = \frac{1}{2}[\{(\alpha_1^2 - \alpha_2^2)(-g^2 + f^2) - 4\alpha_1\alpha_2 fg - (-g'^2 + f'^2)\}\cos(2\alpha_1 t)\cosh(2\alpha_2 t)$$
$$+ \{2\alpha_1\alpha_2 f^2 - 2\alpha_1\alpha_2 g^2 + 2(\alpha_1^2 - \alpha_2^2)gf - 2g'f'\}\sin(2\alpha_1 t)\sinh(2\alpha_2 t)]$$
$$+ \{\alpha_1 fg' + \alpha_1 gf' + \alpha_2(ff' - gg')\}\cos(2\alpha_1 t)\sinh(2\alpha_2 t)$$
$$- \{-\alpha_1 g'g + \alpha_1 f'f - \alpha_2 fg' - \alpha_2 f'g\}\sin(2\alpha_1 t)\cosh(2\alpha_2 t)$$

$$I_{5,2} = \frac{1}{2}[\{-(\alpha_1^2 - \alpha_2^2)(-g^2 + f^2) + 4\alpha_1\alpha_2 fg - g'^2 + f'^2\}\sin(2\alpha_1 t)\sinh(2\alpha_2 t)$$
$$+ \{2\alpha_1\alpha_2(f^2 - g^2) + 2fg(\alpha_1^2 - \alpha_2^2) - 2f'g'\}\cos(2\alpha_1 t)\cosh(2\alpha_2 t)]$$
$$- [\{\alpha_1 f'g + \alpha_1 g'f - \alpha_2 f'f + \alpha_2 gg'\}\sin(2\alpha_1 t)\cosh(2\alpha_2 t)]$$
$$- [\{\alpha_1 f'f - \alpha_1 g'g - \alpha_2 g'f - \alpha_2 f'g\}\cos(2\alpha_1 t)\sinh(2\alpha_2 t)]$$

The Gauss2 method is again used to integrate Equation (36) with stepsize $h = 0.01$ and $n = 10{,}000$ number of steps. The absolute error in the first integrals is calculated as before. The absolute error in integrals $I_{1,1}$, $I_{1,2}$, $I_{3,1}$ and $I_{3,2}$ is plotted in Figures 9–12, respectively, which remains bounded for long time. Similar error behavior is obtained for $I_{2,1}$, $I_{2,2}$, $I_{4,1}$, $I_{4,2}$, $I_{5,1}$, and $I_{5,2}$. The symplectic Gauss2 method is able to preserve all first integrals obtained by performing complex symmetry analysis.

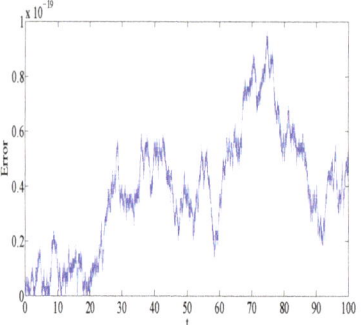

Figure 9. Error in integral $I_{1,1}$.

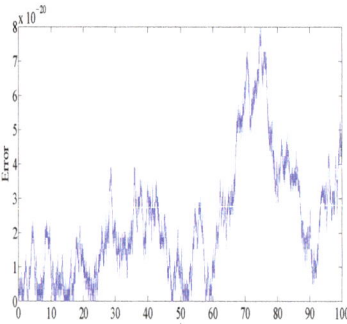

Figure 10. Error in integral $I_{1,2}$.

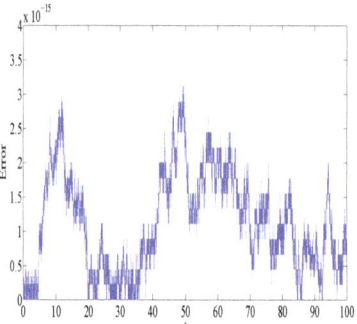

Figure 11. Error in integral $I_{3,1}$.

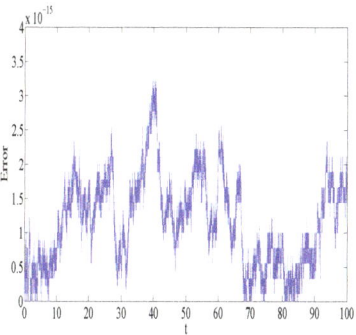

Figure 12. Error in integral $I_{3,2}$.

5. Conclusions

The first integrals of dynamical system $y'' = -k^2 y$ were obtained via the classical Noether approach and the complex symmetry method. The later approach yields invariant energy as a particular example that is stored in both oscillators. Since these first integrals are quadratic in nature, the symplectic Runge–Kutta method, whose construction is also given in this paper, was successfully applied to the system, and numerical preservation of these first integrals was obtained. Interestingly, the numerical method presented in this paper could preserve the energy of the single oscillator as well as the energy stored in the pair of coupled oscillators that arise from the complex Noether approach. The error in the first integrals remained bounded for a long time, which would not have been possible if we have employed nonsymplectic integrators.

Author Contributions: Conceptualization, Y.H.; formal analysis, W.I. and M.U.F.; investigation, W.I.; methodology, Y.H. and M.U.F.; project administration, Y.H.; software, Y.H. and W.I.; supervision, Y.H.; validation, M.U.F.; visualization, W.I.; writing—original draft, W.I.; writing—review and editing, Y.H.

Funding: This research received no external funding.

Conflicts of Interest: The authors declare that there is no conflict of interest.

References

1. Ibragimov, N.H. *Elementary Lie Group Analysis and Ordinary Differential Equations*; John Wiley and Sons: Chichester, UK, 1999.
2. Lie, S. *Theorie der Transformationsgruppen*; Teubner: Leipzig, Germany, 1888.
3. Lie, S. *Vorlesungen über Differentialgleichungen Mit Bekannten Infinitesimalen Transformationen*; Teubner: Leipzig, Germany, 1891.
4. Stephani, H. *Differential Equations: Their Solution Using Symmetries*; Cambridge University Press: Cambridge, UK, 1989.
5. Noether, E. Invariante Variationsprobleme, Nachrichten der Akademie der Wissenschaften in Göttingen. *Mathematisch-Physikalische Klasse* **1918**, *2*, 235–257. English translation in Transport Theory and Statistical Physics **1971**, *1*, 186–207.
6. Ibragimov, N.H.; Kara, A.H.; Mahomed, F.M. Lie-Backlund and Noether symmetries with applications. *Nonlinear Dyn.* **1998**, *15*, 115–136. [CrossRef]
7. Leach, P.G.L. Applications of the Lie theory of extended groups in Hamiltonian mechanics: The oscillator and the Kepler problem. *J. Aust. Math. Soc.* **1981**, *23*, 173–186. [CrossRef]
8. Lutzky, M. Symmetry groups and conserved quantities for the harmonic oscillator. *J. Phys. A Math. Gen.* **1978**, *11*, 249. [CrossRef]
9. Naz, R.; Freire, I.L.; Naeem, I. Comparison of different approaches to construct first integrals for ordinary differential equations. *Abstr. Appl. Anal.* **2014**, *2014*, 1–15. [CrossRef] [PubMed]

10. Farooq, M.U.; Ali, S.; Mahomad, F.M. Two-dimensional systems that arise from the Noether classification of Lagrangians on the line. *Appl. Math. Comput.* **2011**, *217*, 6959–6973. [CrossRef]
11. Farooq, M.U.; Ali, S.; Qadir, A. Invariants of two-dimensional systems via complex Lagrangians with applications. *Commun. Nonlinear Sci. Numer. Simul.* **2011**, *16*, 1804–1810. [CrossRef]
12. Ali, S.; Mahomed, F.M.; Qadir, A. Complex Lie symmetries for variational problems. *J. Nonlinear Math. Phys.* **2008**, *5*, 25–35. [CrossRef]
13. Sanz-Serna, J.M.; Calvo, M.P. Symplectic numerical methods for Hamiltonian problems. *J. Mod. Phys. C* **1993**, *4*, 385–392. [CrossRef]
14. Hairer, E.; Lubich, C.; Wanner, G. *Geometric Numerical Integration: Structure-Preserving Algorithms for Ordinary Differential Equations*, 2nd ed.; Springer: Berlin/Heidelberg, Germany, 2005.
15. Ali, S.; Mahomed, F.M.; Qadir, A. Complex Lie symmetries for scalar second-order ordinary differential equations. *Nonlinear Anal. Real World Appl.* **2009**, *10*, 3335–3344. [CrossRef]
16. Butcher, J.C. A history of Runge-Kutta methods. *Appl. Numer. Math.* **1996**, *20*, 247–260. [CrossRef]
17. Burrage, K.; Butcher, J.C. Stability criteria for implicit Runge-Kutta methods. *SIAM J. Numer. Anal.* **1979**, *16*, 46–57. [CrossRef]
18. Cooper, G.J. Stability of Runge-Kutta methods for trajectory problems. *IMA J. Numer. Anal.* **1987**, *7*, 1–13. [CrossRef]
19. Lasagni, F.M. Canonical Runge-Kutta methods. *ZAMP* **1988**, *39*, 952–953. [CrossRef]
20. Sanz-Serna, J.M.; Calvo, M.P. *Numerical Hamiltonian Problems*, 1st ed.; Chapman and Hall: London, UK, 1994.
21. Sanz-Serna, J.M. Runge-Kutta schemes for Hamiltonian systems. *BIT* **1988**, *28*, 877–883. [CrossRef]
22. Sun, G. A simple way of constructing symplectic Runge-Kutta methods. *J. Comput. Math.* **2000**, *18*, 61–68.
23. Habib, Y. Long-Term Behaviour of G-Symplectic Methods. Ph.D. Thesis, The University of Auckland, Auckland, New Zealand, 2010.

 © 2018 by the authors. Licensee MDPI, Basel, Switzerland. This article is an open access article distributed under the terms and conditions of the Creative Commons Attribution (CC BY) license (http://creativecommons.org/licenses/by/4.0/).

MDPI
St. Alban-Anlage 66
4052 Basel
Switzerland
Tel. +41 61 683 77 34
Fax +41 61 302 89 18
www.mdpi.com

Symmetry Editorial Office
E-mail: symmetry@mdpi.com
www.mdpi.com/journal/symmetry

www.ingramcontent.com/pod-product-compliance
Lightning Source LLC
LaVergne TN
LVHW071948080526
838202LV00064B/6705